Dianwei Qian

Anti-Sway Control for Cranes

Also of Interest

Computational Physics
M. Bestehorn, 2018
ISBN 978-3-11-051513-8, e-ISBN (PDF) 978-3-11-051514-5,
e-ISBN (EPUB) 978-3-11-051521-3

The Fundamentals of Electrical Engineering
F. Hüning, 2014
ISBN 978-3-11-034991-7, e-ISBN (PDF) 978-3-11-034990-0,
e-ISBN (EPUB) 978-3-11-030840-2

Advanced Mechanical Engineering
Series Edited by J. Paulo Davim
ISSN 2367-3796, e-ISBN 2367-380X

Nonlinear Engineering
Editor-in-Chief Reza N. Jazar, 4 Issues per year
ISSN 2192-8010, e-ISSN 2192-8029

Dianwei Qian

Anti-Sway Control for Cranes

Design and Implementation Using MATLAB®

DE GRUYTER

Author
Dr. Dianwei Qian
North China Electric Power University
School of Control and Computer Engin
No. 2 Beinong Road
102206 Beijing, China
dianwei.qian@ncepu.edu.cn

ISBN 978-3-11-051936-5
e-ISBN (PDF) 978-3-11-052010-1
e-ISBN (EPUB) 978-3-11-051943-3
Set-ISBN 978-3-11-052011-8

Library of Congress Cataloging-in-Publication Data
A CIP catalog record for this book has been applied for at the Library of Congress.

Bibliographic information published by the Deutsche Nationalbibliothek
The Deutsche Nationalbibliothek lists this publication in the Deutsche Nationalbibliografie; detailed bibliographic data are available on the Internet at http://dnb.dnb.de.

Typesetting: Integra Software Services Pvt. Ltd.
Printing and binding: CPI books GmbH, Leck
Cover image: Green_beat/iStock/thinkstock
♾ Printed on acid-free paper
Printed in Germany

www.degruyter.com

Dedicated to my wife Weiwei Zhao and our daughters Siqi Qian and Siyuan Qian as well as to my parents Yubao Qian and Lixiang Bai with love and gratitude

Dianwei Qian

Preface

Cranes are widely used in industrial environments due to their heavy payload capabilities. Cranes exist in an enormous variety of forms. According to their primary dynamic properties, cranes can roughly be categorized into three types, such as overhead cranes, boom cranes, and tower cranes. Among the three types of cranes, the location of the suspension cable connection point of overhead cranes can be described by a Cartesian coordinate system. In this Cartesian space, the use of overhead cranes suffers from the natural sway of crane payloads, which is inherently a pendulum-type motion.

Taking safety and positioning accuracy into consideration, the sway has adverse effects on the crane performance. Although skilled operators could eliminate much of the sway by moving cranes in very small increments, such an operation must degrade throughput. The control problem of pendulum-type systems rises up on account of some performance demands, such as small payload sway, high safety, high positioning accuracy, and short transportation time.

The control problem has become one of the central research topics. Most of the previous works in this field consider the payload sway as single-pendulum dynamics. Such consideration may originate from the mechanical structure of overhead cranes. The mechanical structure indicates an overhead crane that consists of trolley, hook, and payload. In the crane, a cable is hanged below the trolley, the hook is supported by the cable, and the payload is attached to the hook by some rigging cables. Assuming that the hook and payload are modeled by a lumped mass point, a simplified crane model could be drawn. The model suggests the crane behaves like a single pendulum. Concerning the single-pendulum crane model, numerous theoretical studies and practical implementations have been carried out.

However, the lumped-mass-point assumption does not always hold true. Under some operating conditions, the hook and payload should be modeled by two distributed mass points rather than a lumped mass point. This distributed-mass-point assumption results in double-pendulum dynamics. The double-pendulum dynamics compound two kinds of pendulum motions with different natural frequencies such that those controllers designed to resist the single-pendulum oscillation are not as effective as expected.

Some control designs and applications have been reported to suppress the double-pendulum oscillations. These research results were previously scattered in many journals and conference proceedings worldwide. This book methodically edits and presents the research results in a unified form.

The book consists of eight chapters. Each chapter concludes with appendices about simulation programs.

Chapter 1 provides some necessary background information about cranes, which are the controlled plant in the book, and consequently, an introduction about the types of cranes is also discussed. Then, a brief historical overview of crane control

DOI 10.1515/9783110520101-202

since 2000 is considered. The history and the state-of-the-art research of both the single-pendulum-type crane control and the double-pendulum-type crane control are reviewed and presented. The chapter then proceeds with some insights for bottleneck issues of control and future research directions.

Chapter 2 investigates the double-pendulum-type overhead crane dynamics via Lagrangian mechanics. The double-pendulum-type motions of cranes include two cases such as lumped-mass payload and distributed-mass payload. For each case, the equations of motion are presented by means of the Euler–Lagrange equations. Since the crane models are presented under some ideal assumptions, some uncertainties associated with real applications are discussed next. In addition, the chapter proceeds with the analysis of oscillation frequencies for double-pendulum-type motions based on the methodology of linearization. The analysis distils the essential properties of double-pendulum-type motions.

Chapter 3 focuses on the anti-sway control problem of double-pendulum-type overhead cranes. The control method is passivity based. First, some basic concepts about passivity are presented. Then, taking double-pendulum-type overhead cranes into account, the passivity property of such systems is investigated. The system stability of the passivity-based control method is proven. Some simulation results are illustrated to support the passivity-based anti-sway control. Finally, all the source codes are drawn.

Chapter 4 investigates the methodology of sliding mode for the anti-sway control problem of double-pendulum-type overhead cranes. The methodology of sliding mode is a class of design tools, which contains some basic design methods and their derivatives. This chapter starts with some basic concepts about sliding mode. Then, two advanced sliding mode methods, such as incremental sliding mode control and hierarchical sliding mode control, are presented for double-pendulum-type overhead cranes. The stability analyses are also given in the sense of Lyapunov. Simulation results are illustrated to support the sliding-mode-based anti-sway control methods. Finally, all the source codes are drawn.

Chapter 5 focuses on the methodology of output feedback for the anti-sway control problem of double-pendulum-type overhead cranes. In practice, the angular velocities of hook and payload are unavailable in most cases. Such a practical issue calls for the output feedback-based control design. This chapter starts with some basic concepts about output feedback. Then, an output feedback-based controller is presented for double-pendulum-type overhead cranes. The stability analysis is also proven in the sense of Lyapunov. Simulation results are illustrated to support the control method. Finally, all the source codes are given.

Chapter 6 explores the methodology of fuzzy logic for the anti-sway control problem of double-pendulum-type overhead cranes. Concerning the control applications, fuzzy-logic-based control (short for fuzzy control) has been proven to be effective for solving unmodeled dynamics. However, fuzzy control needs a fuzzy inference model to determine the final control signal. In real applications, fuzzy control often suffers from the dilemma of exponential rule explosion. Fortunately, a fuzzy inference model called single-input-rule modules (SIRMs) can overcome the dilemma by an ingenious structure. This chapter starts with some basic concepts about fuzzy logic and fuzzy control. Then, a SIRM-based fuzzy controller is presented for double-pendulum-type overhead cranes. Some simulation results are illustrated to support the control method. Finally, all the source codes are listed.

Chapter 7 investigates the methodology of input shaping for the anti-sway control problem of double-pendulum-type overhead cranes. Input shaping is an open-loop control technology. It can implement its control action by intelligently shaping the reference commands. The method can effectively cancel motion-induced oscillations. Inherently, the method is a feed-forward technique. In real applications, it is hard to resist any error without special design. In order to deal with these issues, this chapter starts with some basic concepts about input shaping. One simultaneous input shaper is designed to overcome the double-pendulum oscillations. Meanwhile, one classic convolved input shaper is illustrated as a comparison. Then, the simultaneous input shaper integrated with a SIRM-based fuzzy controller is explored for the transport control problem of double-pendulum-type overhead cranes. Some simulation results are illustrated to support the hybrid control design. Finally, all the source codes are listed.

Chapter 8 presents preliminary results on the anti-sway control development and application for double-pendulum-type overhead cranes, draws some concluding remarks, and summarizes open problems for future research.

Appendix provides a software program that integrates all the methods of the book. The software is user-friendly, designed by graphical user interface. It can help beginners understand and master the anti-sway control of double-pendulum motions.

The book can be used for teaching a graduate-level special-topics course in control applications.

In this book, all the control algorithms and their programs are described separately and classified by the chapter name, which can be run successfully in MATLAB 7.5.0.342 version or in other more advanced versions. If you have questions about algorithms and simulation programs, please feel free to contact Dianwei Qian by E-mail: dianwei.qian@gmail.com.

Dianwei Qian
Beijing

Contents

Acknowledgments

I would like to acknowledge a number of research scientists, postdoctoral fellows, and graduate students who have worked with me at different times and whose Ph.D. dissertations and technical papers contribute to the chapters in this book.

Among these, I am especially thankful to professor Jianqiang Yi and professor Dongbin Zhao of the Institute of Automation, Chinese Academy of Sciences, for their help and support with crane control applications.

I am grateful to Dr. Wei Wang, Dr. Diantong Liu, and Dr. Yinxing Hao for numerous useful discussions about crane control during their stay at the Institute of Automation, Chinese Academy of Sciences.

My graduate students Binbin Yang, Qiang Wang, Hong Liu, and Jinrong Guo from the School of Control and Computer Engineering, North China Electric Power University, contributed to the development of the results in this book while working on their dissertations.

At last but not least, I would like to thank my families for their unconditional dedication, love, and support, and to whom – with my humble gratitude – I dedicate this book.

Dianwei Qian
Beijing

DOI 10.1515/9783110520101-204

1 Introduction

Abstract: This chapter provides some necessary background information about cranes, which are the controlled plant in the book, and consequently, an introduction about the types of cranes is also discussed. Then, a brief historical overview of crane control since 2000 is considered. The history and the state-of-the-art research of both single-pendulum-type crane control and double-pendulum-type crane control are reviewed and presented. The chapter then proceeds with some insights for bottleneck issues of control and future research directions.

Keywords: crane type, crane control, single-pendulum dynamics, double-pendulum dynamics

1.1 Cranes: An Overview

A crane is a machine or device used by humans to move loads, or better known as payload, from one point to another. It is typically equipped with a hoist or wire rope drum, wire ropes or chains, and sheaves that can be used to lift or lower the load or move the load horizontally. The machine can help humans to move loads beyond their capability.

The crane was invented by the Ancient Greeks and Romans for lifting heavy loads; see Figure 1.1 [1]. Such an early crane was operated by human or animal power. The Ancient Greeks were believed to have utilized the machine for constructing tall buildings. Larger cranes were developed later by employing human treadwheels; these cranes aided the lifting of heavier loads. In the High Middle Ages, harbor cranes were introduced to load and unload ships and assist with their construction – some were built into stone towers for extra strength and stability.

Early cranes were constructed from wood. However, with the coming of the Industrial Revolution, cast iron, iron, and steel took over [2]. As far as crane power is concerned, ancient cranes were supplied by the physical exertion of men or animals. In the eighteenth or nineteenth century, steam engines were introduced to power cranes, which remained in use till the late twentieth century. From then on, modern cranes usually use internal combustion engines, electric motors, or hydraulic systems to provide a better lifting capability [3].

Today, millions of cranes serve around the world. Owing to their ability to handle hefty objects, they are widely used in many different environments, such as construction sites, warehouses, and harbors. Cranes exist in an enormous variety of forms. Every crane is tailored to a specific use. Sizes range from the smallest jib cranes, used inside workshops, to the tallest tower cranes, used for constructing tall buildings. Mini cranes are also used for constructing tall buildings in order to facilitate constructions by reaching tight spaces [4].

DOI 10.1515/9783110520101-001

Figure 1.1: Schematic diagram of an ancient Roman crane.

In theory, cranes can be regarded as a class of underactuated Lagrangian systems [5]. First, crane systems are underactuated because they have fewer independent control actuators than degrees of freedom (DOF) to be controlled. For example, in a two-dimensional gantry crane system, both cart position on the girder and payload sway are controlled by a single motor. Second, cranes are classified as Lagrangian systems because their equations of motion can be obtained based on the formulation of Lagrangian mechanics like robotic manipulators. Basically, a crane system consists of a support mechanism, which is a part of its structure, and a hoisting mechanism. The hoisting mechanism of a crane often exhibits an oscillatory behavior due to the underactuation of the system. For this reason, it is important for a crane operation to meet stringent safety requirement.

The oscillatory behavior of cranes originates from the physical structure that all cranes use vertical suspension cables to support the payload. Such a structure thereby creates the possibility of pendulum-like payload oscillation. According to the primary dynamic properties of cranes, the location of the suspension cable connection point can be described by different coordinate systems. Accordingly, cranes can roughly be categorized into three types: *overhead cranes*, *boom cranes*, and *tower cranes* [6].

1.2 Overhead Cranes

The first category is overhead cranes, also named bridge cranes. Displayed in Figure 1.2, such an overhead crane operates in Cartesian space. The trolley moves along a bridge suggested by its name. The motion of the payload is perpendicular to that of the trolley. Overhead cranes that can also travel on a mobile base are often called gantry cranes. Overhead cranes are commonly found in industrial environments such as factories (Figure 1.3(a)) and warehouses (Figure 1.3(b)).

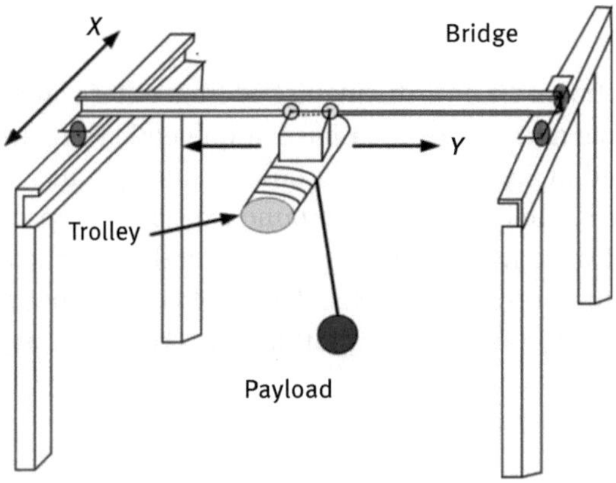

Figure 1.2: Schematic of overhead cranes.

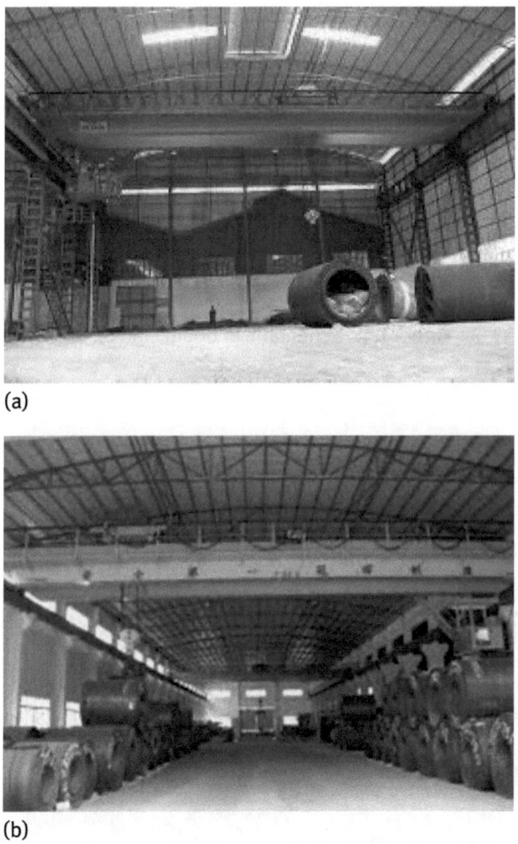

Figure 1.3: Real applications of overhead cranes in China: (a) application in a Chinese factory and (b) application in a Chinese warehouse.

1.3 Boom Cranes

The second category is boom cranes. A boom crane is distinguished from other cranes by its use of a single boom which pivots and rotates on a base at one end. The payload is hoisted from the other. Displayed in Figure 1.4, such a boom crane operates in spherical coordinates. In the coordinates, the boom rotates around axes both perpendicular and parallel to the ground. In Figure 1.4, ψ is the rotation around the vertical Z-axis, and θ is the rotation around the horizontal Y-axis. The payload is supported by a suspension cable at the end of the boom. The platform can be stationary (Figure 1.5(a)) or mobile (Figure 1.5(b)).

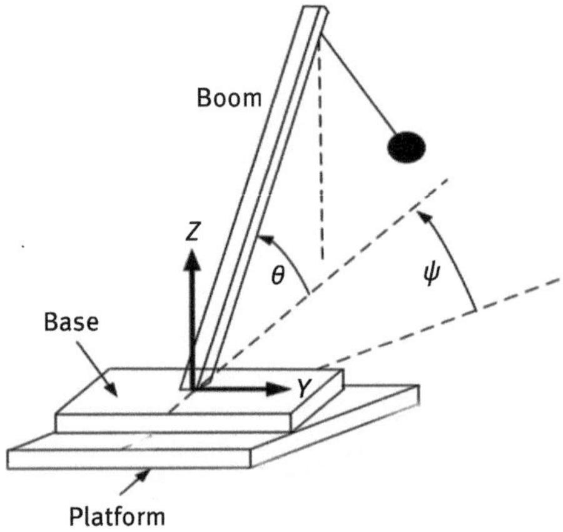

Figure 1.4: Schematic of boom cranes.

(a)

(b)

Figure 1.5: Real applications of boom cranes in China: (a) stationary platform and (b) mobile platform.

Boom cranes are commonly found at building construction sites. Their compact nature also lends well to being mounted on a mobile base. Boom cranes are often mounted on trucks, tracked vehicles, and ships.

1.4 Tower Cranes

The third category is tower cranes. For convenience, the tower crane illustrated in Figure 1.6 can be described by cylindrical coordinates. In the coordinates, the horizontal jib arm can rotate around a vertical tower. The payload is supported by a cable from the trolley and the trolley moves along the jib arm in the radial direction. In Figure 1.6, α and β are the rotational angles around the horizontal plane, θ is the rotation around the vertical Z-axis, and r is the distance between the trolley and the tower.

Tower cranes often give the best combination of height and lifting capacity. They are commonly found in the construction of tall buildings (Figure 1.7) because the type of cranes has the merit of having a small footprint-to-workspace ratio.

1.5 Control Problems

A common characteristic among all cranes is that the payload is supported via an overhead suspension cable. The characteristic results in the same problem of inefficiency caused by payload oscillations. Such payload oscillations make it difficult to

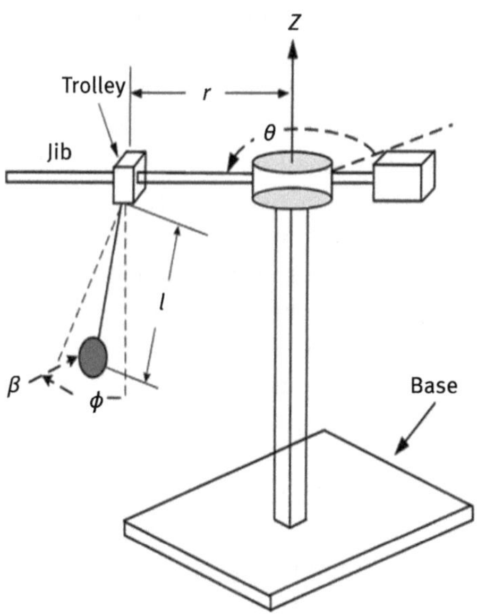

Figure 1.6: Schematic of tower cranes.

Figure 1.7: Real applications of tower cranes in China.

manipulate payloads quickly, accurately, and safely. While this provides the basic functionality of cranes, it also presents several challenges, the primary of which is payload oscillation. Motion of cranes will often translate to large payload oscillations. These payload oscillations have many detrimental effects, including the degradation of payload positioning accuracy, increased task completion time, and decreased safety. Significant research effort has been made into reducing oscillations [7].

To date, the primary method utilized to limit this motion-induced vibration has been to train a skilled operator and to move slowly because the skilled operator could eliminate much of the sway by moving cranes in very small increments. While this does serve to reduce payload oscillations and, thereby, increase safety, it does not provide the most efficient operating condition. Taking safety and positioning accuracy into consideration, payload oscillations have adverse effects on the crane's performance. Although skilled operators could eliminate much of these payload oscillations by moving cranes in very small increments, such an operation must degrade throughput. The control problem of crane systems rises up on account of some performance demands, such as small payload sway, high safety, high positioning accuracy, and short transportation time.

Much of the previous work on crane control has considered single-pendulum dynamics of payload oscillations. Such consideration originates from the mechanical structure of cranes. The mechanical structure of a crane indicates that it contains a trolley, hook, and payload. Consider that a cable is hung below the trolley. The hook is supported by the cable and the payload is attached to the hook by some rigging cables. Assuming that the hook and payload are modeled by a lumped mass point, a simplified crane model could be drawn. This model suggests that payload oscillations behave like a single-pendulum effect. In this case, not only many control approaches presented in the previous work can have good control performance but also experienced crane operators can eliminate much of the sway by causing the payload oscillations during deceleration that cancels the oscillation induced during acceleration.

However, the lumped-mass-point assumption does not always hold true. Under some operating conditions, the hook and payload should be modeled by two distributed mass points rather than a lumped mass point. This distributed-mass-point assumption results in double-pendulum dynamics. The double-pendulum dynamics compound two kinds of pendulum motions with different natural frequencies such that those controllers designed to resist the single-pendulum oscillation are not as effective as expected. Under the double-pendulum oscillations, the manual method of eliminating oscillations becomes very difficult, even for skilled operators. Therefore, *elimination of double-pendulum oscillations by control designs* has become an attractive and emerging benchmark because double-pendulum oscillations are simple enough to permit complete dynamic analyses and experiments, but there exist strong nonlinearity and dynamic coupling.

1.6 Literature Survey

As already mentioned, cranes, regardless of the type, have to move a load from one point to another in the minimum time. Meanwhile, its load should reach its destination without swinging. Usually, a skillful operator is responsible for this task. During the operation, the load is free to swing in a pendulum-like motion. If the swing exceeds a proper limit, it must be damped or the operation must be stopped until the swing dies out. Either option consumes time, which reduces the facility availability. The real problem has motivated many researchers to develop control approaches to automate crane operations. Consider the pendulum motion. The previous work can be divided into two categories: single-pendulum motion and double-pendulum motion.

Control of single-pendulum-type cranes is a mature research area, with a diversity of published control techniques. A systematic overview of these efforts starting in 1961, as well as dynamic crane models, was provided by Abdel-Rahman et al. in 2003 [8]. This book provides a survey of the most recent works on the control of single-pendulum-type cranes.

Compared with a great diversity of control methods for single-pendulum-type cranes, only a few control approaches for double-pendulum-type cranes have been reported till now. Partly, the efforts on controlling double-pendulum-type cranes were investigated by Singhose and Towell until 1998 [9]. Since then, the control of double-pendulum-type cranes has become an attractive topic because the complex double-pendulum dynamics greatly challenge the control design of such kind of cranes.

In the following, crane control is summarized in terms of pendulum dynamics.

1.6.1 Crane Control Based on Single-Pendulum Dynamics

We know that there are two main types of control systems: open-loop control system and closed-loop control system. Accordingly, crane control based on single-pendulum dynamics can also be categorized by this way, namely, open-loop crane control and closed-loop crane control.

1.6.1.1 Open-Loop Crane Control

Open-loop crane control systems are totally independent of the payload sway. Without a doubt, a manual crane control system is just an open-loop one. In order to improve performance, some optimization algorithms are often adopted, such as *optimal trajectory planning* and *command shaping*.

Optimal trajectory planning methods, also named motion planning, seek to eliminate vibration by using preplanned trajectories to move the crane through the workspace. Such kind of methods can avoid the sensing problems of closed-loop crane control methods.

The control problem is typically formulated as a minimum time optimization problem, subject to vibration constraints. Optimal design of the transportation trajectory not only improves the control performance in terms of efficiency or safety but also releases some pressure on control for the sequential strategy.

Lee and his research division solved several motion-planning-based methods to realize the high-performance anti-swing control of overhead cranes, that is, solving the motion planning problem through a kinematic problem [10], designing the motion planning method by means of the model following anti-swing control law [11], and fusing the motion planning method on the sliding mode control (SMC) [12]. Ross and Fahroo [13] proved that the optimal motion planning problem was equivalent to a classic Bolza problem of the calculus of variations and their results were exhibited by the two-dimensional crane model.

To deal with the crane control problems in cluttered work environment, Blajer and Kolodziejczyk [14] first sketched a trajectory by a series of points in the work space and then approximated the trajectory by spline functions. To coordinate multiple construction cranes, the incremental coordination method [15] is employed to plan the motion of two or more cranes, where the kinematics and the geometrical constraints of cranes are taken into consideration. Programming is an effective optimization tool. Several programming methods have been applied to plan crane trajectories, such as dynamic programming [16] and linear programming [17].

Generally speaking, trajectory planning is offline, which makes the method not robust. To deal with the issue, AlBahnassi and Hammad [18] developed a framework for near real-time motion planning of cranes. The framework can assist crane operators to replan safe paths in near real time. Fang and his colleagues in Ref. [19] investigated a kinematic coupling-based offline trajectory planning method for overhead cranes, where the trajectory was tuned by an iterative learning strategy to guarantee accurate trolley positioning. Further, they considered an offline trolley trajectory planning method for underactuated overhead cranes in Ref. [20], where some rigorous geometric analysis was utilized to address the coupling behavior between the actuated trolley motion and the underactuated payload swing.

The primary restriction of the optimal trajectory planning method is that the desired motion of cranes must be known in advance, in addition to the initial conditions of the maneuver. Another major drawback is that generating the optimal profiles can become computationally expensive because of the kinematic coupling behavior between the underactuated payload's swing and the actuated trolley's motion.

Command shaping (usually called input shaping) is another open-loop control technology for cranes, which can dramatically reduce motion-induced payload oscillations by intelligently shaping the reference commands. Unlike the optimal trajectory planning, input shaping is an online method and it can be applied in real time.

To employ the input-shaping method, an input shaper composed of a series of impulses should be designed. These impulses can cancel vibration. That is, vibration caused by a part of the impulses is canceled by vibration caused by other parts of the impulses. The amplitudes and time locations of the impulses are obtained from the system's natural frequencies and damping ratios.

The input-shaping technology is widely used. So far, the method has been implemented on several large bridge cranes at nuclear facilities [21] and a 10-ton crane at Georgia Tech [22]. Input shaping not only achieves crane control directly but also improves crane operator performance [23].

Crane control research in the input-shaping field involves developing shapers to minimize the sway of crane payloads. The use of input shaping dramatically reduces the time to completion of move operations and greatly increases the safety of the crane by allowing for a smaller danger zone around the crane. Singhose and his colleagues also suggested several shaping methods, such as convolved and simultaneous two-mode input shapers [24], expert-system-based shaper [25], and negative-impulse-included shaper [26]. Recently, some novel shaping methods are often reported, such as nonlinear input shaping [27], graphical-approach-based input shaping design [28], and shaping design considering limited-state constraints [29].

Publications concerning crane control by input shaping have resulted in dozens of papers. However, input shapers are located at the feed-forward channel, meaning that they are hard to resist any error without special design. Especially, the drawback becomes deteriorative for some types of double-pendulum cranes because there are severe residual vibrations even for a small modeling error.

1.6.1.2 Closed-Loop Crane Control

Closed-loop crane control aims to find a feedback control law that stabilizes the system in the presence of various uncertainties and external disturbances, which uses information about the current states of the system (e.g., payload swing angle and trolley position) to generate commands that drive the system toward the desired state. Concerning crane control, the control input is the force or torque applied to the trolley in order to suppress oscillations due to the acceleration and deceleration of the trolley. The complexity of the crane control field originates from the crane structure that the number of actuators is less than the DOF to be controlled. Direct applications of many traditional nonlinear control methods also suffer from the underactuated mechanical structure.

In particular, the control of underactuated nonlinear systems has become an important subject of research. Intuitively, the control synthesis for underactuated systems is more complex than that for fully actuated systems. Control of underactuated systems is currently an active field of research due to their broad applications in robotics, land and aerospace vehicles, surface vessels, and crane automation. Since the presence of uncertainties and parameter variations always aroused in underactuated nonlinear systems, the implementation of robust control approach on the systems is appealing.

Recently, many control methods based on feedback linearization, energy, backstepping, sliding mode, and fuzzy logic have been developed for crane control. This section provides a view to highlight some of these popular control methods.

Linear control methods call for the linearized crane model. Reported in Ref. [8], a variety of linear control methods has been applied to crane control practice. Such kind of crane control is mature. However, the crane model is inherently nonlinear, and the linearized method cannot provide sufficient accuracy of information about position error and load swing.

In particular, linear control methods are sensitive to changes in the cable length because they are invariably tuned to counter the effects of the natural frequency of the cable payload at a single cable length. Besides, the uncertain nonlinear factors such as the wind, the hit, and the friction of track will also reduce the performance of linearized crane control systems.

The *partial feedback linearization* method provides a natural global change of coordinates that transform the system into a strict feedback form. Zhang et al. [30] investigated the feasibility of partial feedback linearization for crane control problems; Tuan [31, 32] extended the method to overhead cranes with varying cable length and three-dimensional overhead cranes; Sun and Fang [33] reported their results about the applications of partial feedback linearization with saturation constraint; and Wu et al. [34] designed a novel control method on the basis of partial feedback linearization, where the overall crane system is divided into two subsystems.

Summarily, the advantage of collocated and noncollocated partial feedback linearization methods is a conceptual and structural simplification of control problem, that is, they are always utilized as an initial simplifying step for crane control problems.

Energy-based control. In the sense of Lyapunov's stability theory, a Lyapunov's function candidate is the foundation to analyze the system stability. Pointed out by Maschke et al. [35], it is well known that the total energy is a suitable Lyapunov function to study the stability of the trivial equilibrium of an isolated standard Hamiltonian system. Proven by Karkoub and Zribi [36], cranes are inherently one of isolated standard Hamiltonian systems. Sun and Fang [37] presented some primary results on cranes by energy-based analysis. Some other crane control methods can also be derived from the Lyapunov's stability theory.

Passivity-based control techniques have been used in crane control problems, that is, pendulum-type overhead cranes by Guo et al. [38], convey cranes by Collado et al. [39], and cranes with heavy chains and payload by Thull et al. [40]. Especially, Fang and his colleagues [41] proved the equivalence relation between passivity-based control and a proportional–derivative control concerning the special case of crane control problems.

A nice feature of the passivity-based control design is the physical meaning of the resulting control laws, and the concepts, such as storage energy and dissipation, play a fundamental role in the stability analysis and performance. Stability properties based on the Lyapunov theory can be easily studied for the obtained closed-loop systems. One limitation of passivity-based control is that it is only applicable to the systems with relative degree less than two.

Back-stepping methods can be adapted to design stabilizing controls for a special class of nonlinear dynamical systems. These systems are built from subsystems that radiate out from an irreducible subsystem that can be stabilized using some other method. Because of this recursive structure, a designer can start the design process at the known stable system and "back out" new controllers that progressively stabilize each outer subsystem. The process terminates when the final external control is reached. Hence, this process is entitled back-stepping. In practice, the back-stepping technique often combines other control methods to realize crane control problems, that is, adaptive back-stepping control by Cao et al. [42] and back-stepping SMC by Tsai et al. [43, 44]. However, when the DOF of cranes increases, for instance, double-pendulum-type overhead cranes, the procedure of back-stepping becomes very complicated and implementation of such a control design in practical applications may be unrealistic.

Uncertainty is a common yet intractable issue for crane control. For any control design, there are typical discrepancies between a practical system and its theoretical model because of unmodeled dynamics, parameter uncertainty, and external disturbance.

Adaptive control is the control method used by a controller that must adapt to a controlled system with parameters that vary, or are initially uncertain. An adaptive control system utilizes online identification of either system parameter or controller parameter, which does not need a priori information about the bounds on these uncertain or time-varying parameters. Yang and Yang [45] developed an adaptive control method for three-dimensional cranes, where all of the control system parameters were not needed for the controller design a priori. Recently, some adaptive design approaches against cable length changes [46], external distance [47], and unknown parameters [48] are also reported.

These adaptive control approaches consider their control designs in the sense of Lyapunov according to the traditional adaptive control approaches. Besides, there are still some research results by combining adaptive control and other control methods, including adaptive SMC [49], motion-planning-based adaptive control [50], adaptive sliding mode fuzzy control [51], adaptive coupling control [52], adaptive control based on fuzzy cerebellar model articulation controller [53], to name but a few.

Robust control is a branch of control theory for dealing with bounded system uncertainty. Such a technique is developed under the assumption that uncertain parameters or disturbances are found within some compact set, which aims to achieve robust performance and stability in the presence of bounded modeling errors.

The classic H_∞ loop-shaping robust control, other robust control methods also are developed against the changes of cable length and payload mass, that is, wave-based robust control by Yang and O'Connor [54], pole-placement-based robust control by Uchiyama et al. [55], hybrid control on the basis of proportional–differential control and H_∞ scheme [56], partial-state feedback-based robust control by Uchiyama [57], operator-based robust right coprime factorization design methods by Wen et al. [58–60], linear matrix inequality-based optimal robust control [61], and gain scheduling-based robust control [62], to name but a few.

Although both adaptive and robust controls aim to resist the uncertainties of crane systems, such as cable length and payload mass, the motivations behind the two approaches are different from each other. The idea of robust control can guarantee a designed robust control system to be insensitive to all uncertainties by using a fixed control structure, indicating the robust control law need not be changed if the changes are within the given bounds. On the other hand, the design of an adaptive control system is concerned with the adaptive control law changing itself to suppress all uncertainties. Owing to the nature of these two control structures, robust control is only suitable for dealing with small uncertainty, while adaptive control is suitable for a wide range of parameter variation, but it is sensitive to unstructured uncertainty.

Model predictive control is an advanced control method that has been in use in the process industries such as chemical plants and oil refineries since the 1980s. Deng and Becerra [63] explored the constrained predictive control for cranes. Arnold et al. [64] investigated the application of model predictive control of boom cranes.

Since model predictive controllers rely on dynamic models of the process, most often linear empirical models are obtained by system identification but finding a closed-loop control law for cranes is usually quite complex because of their highly nonlinear equations of motion. As a result, there are not any reports on the aspect of model predictive control of cranes until Van den Broeck et al. [65] presented a time-optimal formulation within the model predictive control framework. Thereafter, model predictive control of crane becomes active again. Some novel approaches on different sides of model predictive control for cranes are reported, that is, solving constraint problems of crane control [66, 67], designing state observers to approximate unmeasurable states [68], and nonlinear model predictive control against crane nonlinearities [69–72], to name but a few.

Suggested by its name, the model predictive control algorithm has the ability to anticipate future events and it can take control actions accordingly. The advantage of model predictive control is the fact that it allows the current timeslot to be optimized, while keeping future timeslots in account. This is achieved by optimizing a finite time horizon, but only implementing the current timeslot. However, main challenges of model predictive control are that the control algorithm highly depends on system model and that the model predictive control algorithm has no guaranteed stability.

Intelligent control is a class of control techniques that use a variety of intelligent computing approaches, such as neural networks, Bayesian probability, fuzzy logic, evolutionary computation, genetic algorithms (GAs), and so on. Among various approaches, evolutionary computation, neural networks, and GAs are often employed to optimize controller parameters and controller structures and they are rare to directly generate control commands. In contrast, fuzzy logic not only has the ability to optimize control systems but also it can directly generate control commands applying to control systems [73]. Consequently, fuzzy logic among these intelligent techniques has also been paid more and more attention.

How to apply the classic theory of fuzzy logic for the crane control problems has been extensively investigated since the 1990s. In the recent two decades, most reports have focused on combination of fuzzy logic and other control techniques in order to improve the crane control performance.

One type can be categorized as fuzzy modeling [74–77], where fuzzy logic is employed to approximate the nonlinear crane model and the final control command is generated by other control methods. The other type can be categorized as fusion of multiple control methods, where one of the control methods is fuzzy logic. Further, the second type can also be subcategorized as fusion of fuzzy logic and evolutionary algorithms [78, 79], combination of fuzzy control and linear quadratic regulator design [80–83], utilization of fuzzy control and proportional–integral–differential control [84–87], and a blend of fuzzy control and adaptive control [88–91]. More recent reports about the fusion of fuzzy control of other control methods, that is, gain scheduling [92], Lyapunov redesign [93], and iterative and evolutionary optimization [94], can be found.

Owing to the nature of strong robustness, the versatile SMC method has received considerable devotion from researchers. Some investigations about crane control problems by sliding mode have been reported in recent years. These contributions can be divided into two types. One type is to investigate various siding mode design methods for cranes. The other type is to fuse other control methods to improve performance of sliding mode-based control systems for cranes.

The first type includes second-order SMC by Bartolini et al. [95] and Vazquez et al. [96–98], high-order SMC by Chen and Saif [99], incremental SMC by Dong et al. [100], terminal SMC by Cao et al. [101], integral SMC by Defoort et al. [102] and Xi and Hesketh [103], hierarchical SMC by Tuan and Lee [104], to name but a few.

The second type is characterized by a blend of SMC and other control methods, which is abundant because of the diversity of control methods. Some representative methods are adaptive sliding mode fuzzy control by Chang et al. [105], fuzzy SMC by Liu et al. [106, 107] and Sun et al. [108], suboptimal integral SMC by Liu et al. [109] and Liu and Li [110], neural network SMC by Wang et al. [111] and Tsai et al. [112], and adaptive SMC by some researchers [113–118].

1.6.2 Crane Control Based on Double-Pendulum Dynamics

Compared with a great diversity of control methods for single-pendulum-type cranes, only a few control approaches for double-pendulum-type cranes have been reported till now, mainly due to their more complicated nonlinearities. In a double-pendulum-type crane, there are three DOFs to be controlled by only one control input, which is very challenging for controller design and analysis. Controlling such a crane system becomes even more challenging especially when no linearization/approximation operations are performed to the nonlinear dynamics.

1.6.2.1 Input-Shaping Technique

For double-pendulum-type cranes, input shaping is the most widely utilized control technique to suppress the double-pendulum swing. The basic principle of this innovative methodology is essentially a finite impulsive response filter, which is determined according to the knowledge of the two linearized frequencies of double-pendulum cranes. Many shaping methods have been reported. Some representative methods are listed as follows.

Fujioka et al. [119] analyzed the robustness of input-shaped model reference control when applied to a double-pendulum crane using a linear single-pendulum crane as the reference model. The robustness of their controller was enhanced to handle both parameter variations and the system order difference between the reference model and the plant.

Boom cranes are used for numerous material-handling and manufacturing processes. They move payloads by slewing their base. The motion induces payload oscillation. The problem of payload oscillation becomes more challenging when the payload exhibits double-pendulum dynamics that produce two varying frequencies of oscillation. Maleki and Singhose [120] and Maleki et al. [121] studied the swing dynamics of such cranes. They also applied input shaping to reduce the two-mode oscillatory dynamics.

Under certain conditions, the payload dynamics of cranes can introduce an additional oscillatory mode that creates a double pendulum on tower cranes. Vaughan et al. [122] presented an analysis of the double-pendulum effect on a tower crane system. They also developed a command generation technique to suppress the oscillatory dynamics with robustness to frequency changes.

The double-pendulum oscillations of crane payloads make it challenging to manipulate payloads quickly, accurately, and safely. Kim and Singhose [123] evaluated an input-shaping control method for reducing double-pendulum oscillations. They illustrated that human operator performance can be much faster and safer with the input-shaping control scheme.

Double-pendulum-type cranes can be treated as rest-to-rest motion of flexible systems with two modes. Sung and Singhose [124] generated limited-state command profiles to suppress the residual vibrations. A great advantage of the command profiles is that they are described by closed-form functions. The commands are designed by first canceling the two modes of flexible vibration and then satisfying a secondary constraint to arrive at the desired target location as quickly as possible.

Daqaq Masoud [125] addressed a nonlinear input-shaping method for double-pendulum-type cranes. The method is based on a nonlinear approximation of the oscillation frequency of the simplified double-pendulum-type crane model. The resulting frequency approximation is used to determine the switching times for a bang-off-bang input-shaping controller.

Masoud and Alhazza [126] developed a frequency-modulation input-shaping technique in order to tune the resonant frequencies of a double-pendulum-type system. By the input-shaping technique, the input commands to a virtual double-pendulum-type crane system can drive a physical one.

Alhazza et al. [127] presented a combined approach that is based on generating shaped commands using the full nonlinear equations of motion combined with the iterative learning control strategy. The combined approach can contribute to the elimination of residual oscillations in rest-to-rest maneuvers of double-pendulum-type cranes.

Masoud et al. [128] explored a hybrid command shaper for double-pendulum overhead cranes. The shaper consists of a primary double-step command shaper complemented by a virtual feedback system. The primary command shaper is designed to eliminate oscillations in a slightly modified version of the crane model with frequencies satisfying the odd integer multiple criterions. The virtual feedback loop is then used to modify the commands of the primary shaper to accommodate the difference between the modified and the original models of the crane.

It is noted that input-shaping methods are open loop, and they are developed and analyzed based on linearized double-pendulum crane models, that is, they assume that the double-pendulum swing angles are small enough during the transferring processes.

1.6.2.2 Feedback Control

In addition to open-loop methods, there are also several feedback control approaches proposed recently [129–141]. Liu et al. [129] analyzed the passivity of double-pendulum-type cranes and proposed a GA-based stable control approach. The stability of the control approach can be guaranteed in the sense of Lyapunov.

In Ref. [130], an incremental SMC scheme for double-pendulum-type system is proposed to achieve asymptotic stability. The asymptotical stability of the entire sliding mode surfaces has been proved theoretically.

Concerning double-pendulum-type overhead cranes, Lahres et al. [131] described a control structure by using the trolley motion and an additional translational actuator at the load suspension unit as system inputs. According to such a control structure, they established a decoupling between position and tilt angle of the load. State-space-based methods were used to design the decoupling control.

To suppress the double-pendulum motions of overhead cranes, the design of simultaneous input-shaping-based fuzzy control is presented in Ref. [132]. Such a design develops the single-input-rule modules (SIRMs) fuzzy controller at first. Then the simultaneous input shaper is adopted to shape the control command generated by the fuzzy controller.

A super-twisting-based nonlinear control law is presented in Ref. [133]. The control law depends on a reduced crane model that can well depict the original double-pendulum dynamics. Such a control law can make the crane follow preset trajectories and eliminate the double-pendulum swing angles.

In Refs. [134, 135], a new energy storage function is adopted to enhance the internal state coupling and then a saturated output feedback controller is designed for underactuated overhead cranes with double-pendulum payload dynamics. The equilibrium point of the closed-loop system is asymptotically stable in the sense of Lyapunov.

Tuan and Lee proposed two discontinuous full-state feedback SMC laws to achieve cart positioning and swing reduction in Ref. [136], and the closed-loop stability is analyzed by neglecting some nonlinear terms.

O'Connor and Habibi in Ref. [137] applied the concept of wave-based control to determine what trolley motion should be used to achieve a reference motion of the load, with minimum swing during complex maneuvers of double-pendulum cargoes. Their presented control design enables the trolley to achieve very good motion control despite the trolley's limited control authority over the complex, flexible double-pendulum cargoes.

Nayfeh and Baumann [138] developed a time-delay feedback control method. The control method depends on a linear approximation specific to the double-pendulum dynamics. For practical operating ranges, the control method undergoes a supercritical bifurcation that helps explain the robustness.

Mar et al. [139] developed a control system that combined input shaping and feedback for double-pendulum systems. The proposed method in Ref. [139] achieves fast point-to-point response similar to open-loop input-shaping control while such a method is able to suppress disturbance-induced residual swing by using the feedback control.

In Ref. [140], an adaptive tracking controller for double-pendulum overhead cranes is developed. The proposed method [140] can guarantee that the trolley tracking error is always within a prior set of boundary conditions and converges to zero rapidly. The asymptotic stability of the closed-loop system's equilibrium point is assured by Lyapunov techniques and Barbalat's lemma.

Qian et al. [141] investigated the SIRMs fuzzy method for transport control of double-pendulum-type crane systems. The controller in Ref. [141] includes six SIRMs that are dynamically weighted. GA is adopted to tune some parameters of the controller.

1.7 Conclusions

Crane control based on single-pendulum dynamics has been extensively researched. Both open-loop and closed-loop methods have been developed to deal with this control problem. Recently, crane control based on double-pendulum dynamics is emerging and has been paid more and more attention.

Some several control methods have been reported to solve the control problem of double-pendulum cranes. These control methods previously are scattered in many journals and conferences worldwide. This book methodically edits and presents some representative control methods in a unified form. The theoretical analysis, simulation results, and source codes of the representative control methods will be released and illustrated.

References

[1] Kurrer KE. The History of the theory of structures: From arch analysis to computational mechanics. Berlin: Ernst & Sohn, 2008.

[2] Lancaster L. Building Trajan's column. Am J Archaeol, 1999, 103(3): 419–439.

[3] Matthies A. Medieval treadwheels. Artists' views of building construction. Technol Cult, 1992, 33(3): 510–547.

[4] Dickie DE. Crane handbook. London: Butterworth-Heinemann, 2013.

[5] Wang W, Yi J, Zhao D, Liu D. Design of a stable sliding-mode controller for a class of second-order underactuated systems. IEE Proc-Control Theory Appl, 2004, 151(6): 683–690.

[6] Vaughan J, Kim D, Singhose W. Control of tower cranes with double-pendulum payload dynamics. IEEE Trans Control Syst Technol, 2010, 18(6): 1345–1358.

[7] Kharola A, Patil PP. Automated control and optimisation of overhead cranes. Int J Manuf Mater Mech Eng 2017, 7(3): 41–68.

[8] Abdel-Rahman EM, Nayfeh AH, Masoud ZN. Dynamics and control of cranes: A review. J Vib Control, 2003, 9(7): 863–908.

[9] Singhose WE, Towell ST. Double-pendulum gantry crane dynamics and control. In: Proceedings of IEEE Conference on Control Applications, Trieste, Italy, 1998, pp. 1205–1209.

[10] Lee HH. A new motion-planning scheme for overhead cranes with high-speed hoisting. J Dyn Syst Meas Control Trans ASME, 2004, 126(2): 359–364.

[11] Lee HH. Motion planning for three-dimensional overhead cranes with high-speed load hoisting. Int J Control, 2005, 78(12): 875–886.

[12] Lee HH, Liang Y, Segura D. A sliding-mode antiswing trajectory control for overhead cranes with high-speed load hoisting. J Dyn Syst Meas Control Trans ASME, 2006, 128(4): 842–845.

[13] Ross IM, Fahroo F. Pseudospectral methods for optimal motion planning of differentially flat systems. IEEE Trans Autom Control, 2004, 49(8): 1410–1413.

[14] Blajer W, Kolodziejczyk K. Motion planning and control of gantry cranes in cluttered work environment. IET Control Theory Appl, 2007, 1(5): 1370–1379.

[15] Kang SC, Miranda E. Computational methods for coordinating multiple construction cranes. J Comput Civil Eng, 2008, 22(4): 252–263.

[16] Zameroski D, Starr G, Wood J, Lumia R. Rapid swing-free transport of non-linear payloads using dynamic programming. J Dyn Syst Meas Control Trans ASME, 2008, 130(4), DOI: 10.1115/1.2936384.

[17] Da Cruz JJ, Leonardi F. Minimum-time anti-swing motion planning of cranes using linear programming. Optim Control Appl Meth, 2013, 34(2): 191–201.

[18] AlBahnassi H, Hammad A. Near real-time motion planning and simulation of cranes in construction: Framework and system architecture. J Comput Civil Eng, 2012, 26(1): 54–63.

[19] Sun N, Fang YC, Zhang YD, Ma BJ. A novel kinematic coupling-based trajectory planning method for overhead cranes. IEEE ASME Trans Mechatron, 2012, 17(1): 166–173.

[20] Sun N, Fang Y, Zhang X, Yuan Y. Transportation task-oriented trajectory planning for underactuated overhead cranes using geometric analysis. IET Control Theory Appl, 2012, 6(10): 1410–1423.

[21] Singer N, Singhose W, Kriikku E. An input shaping controller enabling cranes to move without sway. In: Proceedings of 7th Topical Meeting on Robotics and Remote Systems, Augusta, GA, 1997, pp. 225–231.

[22] Sorensen K, Singhose W, Dickerson S. A controller enabling precise positioning and sway reduction in bridge and gantry cranes. Control Eng Pract, 2007, 15(7): 825–837.

[23] Khalid A, Huey J, Singhose W, Lawrence J, Frakes D. Human operator performance testing using an input-shaped bridge crane. J Dyn Syst Meas Control Trans ASME, 2006, 128(4): 835–841.

[24] Singhose W, Crain E, Seering W. Convolved and simultaneous two-mode input shapers. IEE Proc Control Theory Appl, 1997, 144(6): 515–520.

[25] French L, Singhouse W, Seering W. An expert system for the design of input shapers. In: Proceedings of the 1999 IEEE International Conference on Control Applications, Kohala Coast-Island, USA, 1999, pp. 713–718.

[26] Singhose W, Mills B, Seering W. Vibration reduction with specified-swing input shapers. In: Proceedings of IEEE International Conference on Control Applications, Kohala Coast-Island, USA, 1999, pp. 533–538.

[27] Daqaq MF, Masoud ZN. Nonlinear input-shaping controller for quay-side container cranes. Nonlinear Dyn, 2006, 45(1–2): 149–170.

[28] Masoud ZN, Daqaq MF. A graphical approach to input-shaping control design for container cranes with hoist. IEEE Trans Control Syst Technol, 2006, 14(6): 1070–1077.

[29] Sung YG, Singhose W. Limited-state commands for systems with two flexible modes. Mechatronics, 2009, 19(5): 780–787.

[30] Zhang X, Gao B, Chen H. Nonlinear controller for a gantry crane based on partial feedback linearization. In: Proceedings of International Conference on Control and Automation, Budapest, Hungary, 2005, pp. 1074–1078.

[31] Tuan L, Janchiv A, Kim GH, Lee SG. Feedback linearization control of overhead cranes with varying cable length. In: Proceedings of International Conference on Control, Automation and Systems, Gyeonggi-do, Korea, 2011, pp. 906–911.

[32] Tuan L, Kim GH, Lee SG. Partial feedback linearization control of the three dimensional overhead crane. In: Proceedings of IEEE International Conference on Automation Science and Engineering, Seoul, Korea, 2012, pp. 1198–1203.

[33] Sun N, Fang YC. A partially saturated nonlinear controller for overhead cranes with experimental implementation. In: Proceedings of IEEE International Conference on Robotics and Automation, Karlsruhe, Germany, 2013, pp. 4473–4478.

[34] Wu XQ, He XX, Sun N, Fang YC. A novel anti-swing control method for 3-D overhead cranes. In: Proceedings of American Control Conference, Portland OR, USA, 2014, pp. 2821–2826.

[35] Maschke B, Ortega R, Van der Schaft AJ. Energy-based Lyapunov functions for forced Hamiltonian systems with dissipation. IEEE Trans Autom Control, 2000, 45(8): 1498–1502.

[36] Karkoub MA, Zribi M. Modelling and energy based nonlinear control of crane lifters. IEE Proc Control Theory Appl, 2002, 149(3): 209–216.

[37] Sun N, Fang YC. New energy analytical results for the regulation of underactuated overhead cranes: An end-effector motion-based approach. IEEE Trans Ind Electron, 2012, 59(12): 4723–4734.

[38] Guo W, Liu D, Yi J, Zhao D. Passivity-based-control for double-pendulum-type overhead cranes. In: Proceedings of IEEE Region 10 Annual International Conference, Chiang Mai, Thailand, 2004, pp. 546–549.

[39] Collado J, Lozano R, Fantoni I. Control of convey-crane based on passivity. In: Proceedings of American Control Conference, Chicago IL, USA, 2000, pp. 1260–1264.

[40] Thull D, Wild D, Kugi A. Application of a combined flatness- and passivity-based control concept to a crane with heavy chains and payload. In: Proceedings of IEEE International Conference on Control Applications, Munich, Germany, 2006, pp. 656–661.

[41] Fang Y, Dixon WE, Dawson DM, Zergeroglu E. Nonlinear coupling control laws for an underactuated overhead crane system. IEEE ASME Trans Mechatron, 2003, 8(3): 418–423.

[42] Cao LZ, Wang HW, Niu C, Wei SB. Adaptive backstepping control of crane hoisting system. In: Proceedings of IEEE International Conference on Automation and Logistics, Qingdao, China, 2007, pp. 245–248.

[43] Tsai CC, Wu HL, Chuang KH. Backstepping aggregated sliding-mode motion control for automatic 3D overhead cranes. In: Proceedings of IEEE/ASME International Conference on Advanced Intelligent Mechatronics, Kaohsiung, Taiwan, 2012, pp. 849–854.

[44] Tsai CC, Wu HL, Chuang KH. Intelligent sliding-mode motion control using fuzzy wavelet networks for automatic 3D overhead cranes. In: Proceedings of 51st Annual Conference of the Society of Instrument and Control Engineers of Japan, Akita, Japan, 2012, pp. 1256–1261.

[45] Yang JH, Yang KS. Adaptive control for 3-D overhead crane systems. In: Proceedings of American Control Conference, Minneapolis MN, USA, 2006, pp. 1832–1837.

[46] Messineo S, Serrani A. Offshore crane control based on adaptive external models. In: Proceedings of American Control Conference, Seattle WA, USA, 2008, pp. 2498–2503.

[47] Sun N, Fang Y, Chen H. Adaptive control of underactuated crane systems subject to bridge length limitation and parametric uncertainties. In: Proceedings of the 33rd Chinese Control Conference, Nanjing, China, 2014, pp. 3568–3573.
[48] He W, Zhang S, Ge SS. Adaptive control of a flexible crane system with the boundary output constraint. IEEE Trans. Ind. Electron, 2014, 61(8): 4126–4133.
[49] Ngo QH, Hong KS. Adaptive sliding mode control of container cranes. IET Control Theory Appl, 2010, 6(5): 662–668.
[50] Fang Y, Ma B, Wang P, Zhang X. A motion planning-based adaptive control method for an underactuated crane system. IEEE Trans Control Syst Technol, 2012, 20(1): 241–248.
[51] Liu D, Yi J, Zhao D, Wang W. Adaptive sliding mode fuzzy control for a two-dimensional overhead crane. Mechatronics, 2005, 15(5): 505–522.
[52] Yang JH, Yang KS. Adaptive coupling control for overhead crane systems. Mechatronics, 2007, 17(2): 143–152.
[53] Yu W, Moreno-Armendariz MA, Rodriguez FO. Stable adaptive compensation with fuzzy CMAC for an overhead crane. Inf Sci, 2011, 181(21): 4895–4907.
[54] Yang TW, O'Connor WJ. Wave based robust control of a crane sys-tem. In: Proceedings of 2006 IEEE/RSJ International Conference on Intelligent Robots and Systems, Beijing, China, 2006, pp. 2724–2729.
[55] Uchiyama N, Takagi S, Sano S. Robust control of rotary cranes based on pole placement approach. In: Proceedings of the 9th IEEE International Workshop on Advanced Motion Control, Istanbul, Turkey, 2006, pp. 647–652.
[56] Toda M. A unified approach to robust control of flexible mechanical systems. In: Proceedings of the 46th IEEE Conference on Decision and Control, New Orleans LA, USA, 2007, pp. 5787–5793.
[57] Uchiyama N. Robust control of rotary crane by partial-state feedback with integrator. Mechatronics, 2009, 19(8): 1294–1302.
[58] Wen SJ, Deng MC, Ohno Y, Wang DY. Operator-based robust right coprime factorization design for planar gantry crane. In: Proceedings of 2010 International Conference on Mechatronics and Automation, Xi'an, China, 2011, pp. 1–5.
[59] Wen SJ, Deng MC, Wang DY. Operator-based robust nonlinear control for a crane system with constraint inputs. In: Proceedings of the 30th Chinese Control Conference, Yantai, China, 2011, pp. 6109–6114.
[60] Wen SJ, Deng MC, Inoue A. Operator-based robust non-linear control for gantry crane system with soft measurement of swing angle. Int J Model Identif Control, 2012, 16(1): 86–96.
[61] Sano S, Ouyang H, Uchiyama N. Robust control of rotary cranes under rope length variance via LMI optimization. In: Proceedings of IEEE International Conference on Industrial Technology, Busan, Korea, 2012, pp. 846–851.
[62] Ushida Y, Narita M, Ushiro T, Chen G, Takami I. Robust control for crane considering all varying parameters in the dynamics. In: Proceedings of the 11th IEEE International Conference on Control and Automation, Tai-chung, Taiwan, 2014, pp. 785–790.
[63] Deng JM, Becerra VM. Application of constrained predictive control on a 3D crane system. In: Proceedings of IEEE Conference on Robotics, Automation and Mechatronics, Singapore City, Singapore, 2004, pp. 583–587.
[64] Arnold E, Sawodny O, Neupert J, Schneider K. Anti-sway system for boom cranes based on a model predictive control approach. In: Proceedings of IEEE International Conference on Mechatronics and Automation, Niagara Falls, Canada, 2005, pp. 1533–1538.
[65] Van den Broeck L, Diehl M, Swevers J. A model predictive control approach for time optimal point-to-point motion control. Mechatronics, 2011, 21(7): 1203–1212.

[66] Kapernick B, Graichen K. Model predictive control of an overhead crane using constraint substitution. In: Proceedings of American Control Conference, Washington DC, USA, 2013, pp. 3973–3978.

[67] Schaper U, Arnold E, Sawodny O, Schneider K. Constrained real-time model-predictive reference trajectory planning for rotary cranes. In: Proceedings of IEEE/ASME International Conference on Advanced Intelligent Mechatronics, Wollongong, Australia, 2013, pp. 680–685.

[68] Bauer D, Schaper U, Schneider K, Sawodny O. Observer design and flatness-based feedforward control with model predictive trajectory planning of a crane rotator. In: Proceedings of American Control Conference, Portland OR, USA, 2014, pp. 4020–4025.

[69] Barisa T, Bartulovic M, Zuzic G, Iles S, Matusko J, Kolonic F. Non-linear predictive control of a tower crane using reference shaping approach. In: Proceedings of the 16th International Power Electronics and Motion Control Conference and Exposition, Antalya, Turkey, 2014, pp. 872–876.

[70] Iles S, Matusko J, Kolonic F. Real-time predictive control of 3D tower crane. In: Proceedings of IEEE 23rd International Symposium on Industrial Electronics, Istanbul, Turkey, 2014, pp. 224–230.

[71] Khatamianfar A, Savkin AV. A new tracking control approach for 3D over-head crane systems using model predictive control. In: Proceedings of European Control Conference, Strasbourg, France, 2014, pp. 796–801.

[72] Bock M, Kugi A. Real-time nonlinear model predictive path-following control of a laboratory tower crane. IEEE Trans Control Syst Technol, 2014, 22(4): 1461–1473.

[73] Michels K, Klawonn F, Kruse R, Numberger A. Fuzzy control: Fundamentals, stability and design of fuzzy controllers. New York: Springer-Verlag, 2006.

[74] Al-Humaidi HM, Hadipriono Tan F. Mobile crane safe operation approach to prevent electrocution using fuzzy-set logic models. Adv Eng Softw, 2009, 40(8): 686–696.

[75] Chen YJ, Wang WJ, Chang CL. Guaranteed cost control for an over-head crane with practical constraints: Fuzzy descriptor system approach. Eng Appl Artif Intell, 2009, 22, pp. 639–645.

[76] Smoczek J, Szpytko J. Evolutionary algorithm-based design of a fuzzy TBF predictive model and TSK fuzzy anti-sway crane control system. Eng Appl Artif Intell, 2014, 28: 190–200.

[77] Precup RE, Filip HI, Rădac MB, et al. Online identification of evolving Takagi–Sugeno–Kang fuzzy models for crane systems. Appl Soft Comput J, 2014, 24: 1155–1163.

[78] Liu D, Yi J, Tan M. Proposal of GA-based two-stage fuzzy control of over-head crane. In: Proceedings of IEEE Region 10 Conference on Computer, Communications, Control and Power Engineering, Beijing, China, 2002, pp. 1721–1724.

[79] Petrenko YN, Alavi SE. Fuzzy logic and genetic algorithm technique for non-liner system of overhead crane. In: Proceedings of 2010 IEEE Region 8 International Conference on Computational Technologies in Electrical and Electronics Engineering, Irkutsk Listvyanka, Russia, 2010, pp. 848–851.

[80] Adeli M, Zarabadi SH, Zarabadipour H, Aliyari Shoorehdeli M. Design of a parallel distributed fuzzy LQR controller for double-pendulum-type overhead cranes. In: Proceedings of IEEE International Conference on Control System, Computing and Engineering, Penang, Malaysia, 2011, pp. 62–67.

[81] Adeli M, Zarabadipour H, Zarabadi SH, Aliyari Shoorehdeli M. Anti-swing control for a double-pendulum-type overhead crane via parallel distributed fuzzy LQR controller combined with genetic fuzzy rule set selection. In: Proceedings of IEEE International Conference on Control System, Computing and Engineering, Penang, Malaysia, 2011, pp. 306–311.

[82] Adeli M, Zarabadipour H, Shoorehdeli MA. Crane control via parallel distributed fuzzy LQR controller using genetic fuzzy rule selection. In: Proceedings of International Conference on Control, Instrumentation and Automation, Shiraz, Iran, 2011, pp. 390–395.
[83] Adeli M, Zarabadipour H, Shoorehdeli MA. Anti-swing control of a double-pendulum-type overhead crane using parallel distributed fuzzy LQR controller. In: Proceeding of International Conference on Control, Instrumentation and Automation, Shiraz, Iran, 2011, pp. 401–406.
[84] Solihin MI, Wahyudi A. Fuzzy-tuned PID control design for automatic gantry crane. In: Proceedings of International Conference on Intelligent and Advanced Systems, Kuala Lumpur, Malaysia, 2007, pp. 1092–1097.
[85] Yang Y. Fuzzy-PI damping control for hydraulic crane tip. In: Proceedings of 5th International Conference on Fuzzy Systems and Knowledge Discovery, Jinan, China, 2008, pp. 75–79.
[86] Lin F, Zhang X, Zhai X, Huang H. The application study of heave compensation control based on motion prediction and fuzzy-PID for intelligence crane. In: Proceedings of International Conference on Mechatronics and Automation, Beijing, China, 2011, pp. 2370–2374.
[87] Liu C, Zhao H, Cui Y. Research on application of fuzzy adaptive PID controller in bridge crane control system. In: Proceedings of IEEE International Conference on Software Engineering and Service Science, Beijing, China, 2014, pp. 971–974.
[88] Chang CY. Adaptive fuzzy controller of the overhead cranes with nonlinear disturbance. IEEE Trans Ind Inf, 2007, 3(2): 164–172.
[89] Park MS, Chwa D, Hong SK. Adaptive fuzzy nonlinear anti-sway trajectory tracking control of uncertain overhead cranes with high-speed load hoisting motion. In: Proceedings of International Conference on Control, Automation and Systems, Seoul, South Korea, 2007, pp. 2886–2891.
[90] Pal AK, Mudi RK. An adaptive fuzzy controller for overhead crane. In: Proceedings of IEEE International Conference on Advanced Communication Control and Computing Technologies, Ramanathapuram, India, 2012, pp. 300–304.
[91] Smoczek J, Szpytko J. Fuzzy logic-based adaptive control system proto-typing for laboratory scaled overhead crane. In: Proceedings of 18th International Conference on Methods and Models in Automation and Robotics, Międzyzdroje, Poland, 2013, pp. 92–97.
[92] Smoczek J, Szpytko J. Design of gain scheduling anti-sway crane controller using genetic fuzzy system. In: Proceedings of 17th International Conference on Methods and Models in Automation and Robotics, 2012, pp. 573–578.
[93] Sharkawy AB, Moustafa K, El-Awady H, et al. Control of overhead crane based on Lyapunov approach to fuzzy controller synthesis. In: Proceedings of 23rd International Conference on Robotics in Alpe-Adria-Danube Region, Smolenice Castle, Slovakia, 2014, pp. 1–6.
[94] Smoczek J, Szpytko J. Iterative and evolutionary optimization for interval analysis-based designing a fuzzy controller for a planar crane model. In: Proceedings of 19th International Conference on Methods and Models in Automation and Robotics, Międzyzdroje, Poland, 2014, pp. 258–263.
[95] Bartolini G, Pisano A, Usai E. Second-order sliding-mode control of container cranes. Automatica, 2002, 38(10): 1783–1790.
[96] Vazquez C, Collado J, Fridman L. On the second order sliding mode control of a parametrically excited overhead-crane. In: Proceedings of American Control Conference, Montréal, Canada, 2012, pp. 6288–6293.
[97] Vazquez C, Fridman L, Collado J. Second order sliding mode control of an overhead-crane in the presence of external perturbations. In: Proceedings of IEEE 52nd Annual Conference on Decision and Control, Florence, Italy, 2013, pp. 2876–2880.

[98] Vazquez C, Aranovskiy S, Freidovich L, Fridman L. Second order sliding mode control of a mobile hydraulic crane. In: Proceedings of IEEE 53rd annual Conference on Decision and Control, Los Angeles CA, USA, 2014, pp. 5530–5535.

[99] Chen WT, Saif M. Output feedback controller design for a class of MIMO nonlinear systems using high-order sliding-mode differentiators with application to a laboratory 3-D crane. IEEE Trans Ind Electron, 2008, 55(11): 3985–3997.

[100] Dong Y, Wang Z, Feng Z, Cheng J. Incremental sliding mode control for double-pendulum-type overhead crane system. In: Proceedings of 27th Chinese Control Conference, Kunming, China, 2008, pp. 368–371.

[101] Cao L, Li XB, Lou FP, et al. Controller design and simulation of the crane based on non-singular terminal sliding mode method. In: Proceedings of International Conference on Logistics Systems and Intelligent Management, Harbin, China, 2010, pp. 484–487.

[102] Defoort M, Maneeratanaporn J, Murakami T. Integral sliding mode anti-sway control of an underactuated overhead crane system. In: Proceedings of 9th France-Japan & 7th Europe-Asia Congress on Mechatronics and Research and Education in Mechatronics, Paris, France, 2012, pp. 71–77.

[103] Xi Z, Hesketh T. Discrete time integral sliding mode control for over-head crane with uncertainties. IET Control Theory Appl, 2010, 4(10): 2071–2081.

[104] Tuan LA, Lee SG. Sliding mode controls of double-pendulum crane systems. J Mech Sci Technol, 2013, 27(6): 1863–1873.

[105] Chang CY, Hsu KC, Chiang KH, Huang GE. An enhanced adaptive sliding mode fuzzy control for positioning and anti-swing control of the overhead crane system. In: Proceedings of IEEE International Conference on Systems, Man and Cybernetics, Taipei, Taiwan, 2006, pp. 992–997.

[106] Liu DT, Yi JQ, Zhao DB. Fuzzy tuning sliding mode control of transporting for an overhead crane. In: Proceedings of International Conference on Machine Learning and Cybernetics, Xi'an, China, 2003, pp. 2541–2546.

[107] Liu DT, Yi JQ, Zhao DB, Wang W. Swing-free transporting of two-dimensional overhead crane using sliding mode fuzzy control. In: Proceedings of American Control Conference, Boston MA, USA, 2004, pp. 1764–1769.

[108] Sun H, Chen ZM, Meng WJ. Fuzzy sliding mode anti-swing control for tower crane base on time-delayed filter. In: Proceedings of 24th Chinese Control and Decision Conference, Taiyuan, China, 2012, pp. 2205–2210.

[109] Liu R, Li S, Chen X. An optimal integral sliding mode control design based on pseudospectral method for overhead crane systems: In: Proceedings of 32nd Chinese Control Conference, Xi'an, China, 2013, pp. 2195–2200.

[110] Liu R, Li S. Suboptimal integral sliding mode controller design for a class of affine systems. J Optim Theory Appl, 2014, 161(3): 877–904.

[111] Wang W, Yi JQ, Zhao DB, Liu XJ. Incremental neural network sliding mode controller for an overhead crane. In: Proceedings of International Conference on Intelligent Mechatronics and Automation, Chengdu, China, 2004, pp. 166–171.

[112] Tsai CC, Wu HL, Chuang KH. Intelligent sliding-mode motion control using fuzzy wavelet networks for automatic 3D overhead cranes. In: Proceedings of 51st Annual Conference of the Society of Instrument and Control Engineers of Japan, Akita, Japan, 2012, pp. 1256–1261.

[113] Tuan LA, Moon SC, Kim DH, Lee SG. Adaptive sliding mode control of three dimensional overhead cranes. In: Proceedings of IEEE International Conference on Cyber Technology in Automation, Control, and Intelligent Systems, Bangkok, Thailand, 2012, pp. 354–359.

[114] Tuan LA, Moon SC, Lee WG, Lee SG. Adaptive sliding mode control of overhead cranes with varying cable length. J Mech Sci Technol, 2013, 27(3): 885–893.

[115] Qian DW, Tong SW, Yi JQ. Adaptive control based on incremental hierarchical sliding mode for overhead crane systems. Appl Math Inform Sci, 2013, 7(4): 1359–1364.
[116] Tuan LA, Lee SG, Nho LC, Kim DH. Model reference adaptive sliding mode control for three dimensional overhead cranes. Int J Precis Eng Manuf, 2013, 14(8): 1329–1338.
[117] Park MS, Chwa D, Eom M. Adaptive sliding-mode antisway control of uncertain overhead cranes with high-speed hoisting motion. IEEE Trans Fuzzy Syst, 2014, 22(5): 1262–1271.
[118] Dhahri S, Hmida FB, Sellami A, Gossa M. Actuator fault reconstruction for linear uncertain systems using sliding mode observer. In: Proceedings of 3rd International Conference on Signals, Circuits and Systems, Medenine, Tunisia, 2009, pp. 1–6.
[119] Fujioka D, Shah M, Singhose W. Robustness analysis of input-shaped model reference control on a double-pendulum crane. In: Proceedings of the American Control Conference, Chicago, USA, 2015, pp. 2561–2566
[120] Maleki E, Singhose W. Swing dynamics and input-shaping control of human-operated double-pendulum boom cranes. J Comput Nonlinear Dyn, 2012, 7(3). DOI: 10.1115/1.4005933
[121] Maleki E, Singhose W, Srinivasan S. Positioning and control of boom crane luffing with double-pendulum payloads. In: Proceedings of IEEE International Conference on Control Applications, Yokohama, Japan, 2010, pp. 1319–1324.
[122] Vaughan J, Kim D, Singhose W. Control of tower cranes with double-pendulum payload dynamics. IEEE Trans Control Syst Technol, 2010, 18(6): 1345–1358.
[123] Kim D, Singhose W. Performance studies of human operators driving double-pendulum bridge cranes. Control Eng Practice, 2010, 18(6): 567–576.
[124] Sung YG, Singhose WE. Limited-state commands for systems with two flexible modes. Mechatronics, 2009, 19(5): 780–787.
[125] Daqaq MF, Masoud ZN. Nonlinear input-shaping controller for quay-side container cranes. Nonlinear Dyn, 2006, 45(1–2): 149–170.
[126] Masoud ZN, Alhazza KA. Frequency-modulation input shaping control of double-pendulum overhead cranes. J Dyn Syst Meas Control-Trans. ASME, 2014, 136(2): DOI: 10.1115/1.4025796
[127] Alhazza KA, Hasan AM, Alghanim KA, Masoud ZN. An iterative learning control technique for point-to-point maneuvers applied on an overhead crane. Shock Vib, 2014, DOI: 10.1155/2014/261509
[128] Masoud Z, Alhazza K, Abu-Nada E, Majeed M. A hybrid command-shaper for double-pendulum overhead cranes. J Vib Control, 2014, 20(1): 24–37.
[129] Liu DT, Guo WP, Yi JQ. Dynamics and GA-based stable control for a class of underactuated mechanical systems. Int J Control Autom Syst, 2008, 6(1): 35–43.
[130] Dong YY, Wang ZH, Feng ZQ, Cheng J. Incremental sliding mode control for double-pendulum-type overhead crane system. In: Proceedings of the 27th Chinese Control Conference, Kunming, China, 2008, pp. 368–371.
[131] Lahres S, Aschemann H, Sawodny O, Hofer EP. Crane automation by decoupling control of a double pendulum using two translational actuators. In: Proceedings of the American Control Conference, Chicago, USA 2000, pp. 1052–1056.
[132] Qian D, Tong S, Yang B, Lee S. Design of simultaneous input-shaping-based SIRMs fuzzy control for double-pendulum-type overhead cranes. Bull Pol Acad Sci-Tech Sci, 2015, 63(4): 887–896.
[133] Sun N, Fang YC, Chen H, Fu YM. Super-twisting-based anti-swing control for under-actuated double pendulum cranes. In: Proceedings of IEEE/ASME International Conference on Advanced Intelligent Mechatronics, Busan, South Korea, 2015, pp. 749–754.
[134] Sun N, Fang YC, Chen H, Lu B. Energy-based control of double pendulum cranes. In: Proceedings of IEEE Annual International Conference on Cyber Technology in Automation Control and Intelligent Systems, Shenyang, China 2015, pp. 258–263.

[135] Sun N, Fang YC, Chen H, Lu B. Amplitude-saturated nonlinear output feedback antiswing control for underactuated cranes with double-pendulum cargo dynamics. IEEE Trans Ind Electron, 2017, 64(3): 2135–2146.
[136] Tuan LA, Lee SG, Sliding mode controls of double-pendulum crane systems. J Mech Sci Technol, 2013, 27(6): 1863–1873.
[137] O'Connor W, Habibi H, Gantry crane control of a double-pendulum, distributed-mass load, using mechanical wave concepts. Mech Sci, 2013, 4(2): 251–261.
[138] Nayfeh NA, Baumann WT. Nonlinear analysis of time-delay position feedback control of container cranes. Nonlinear Dyn, 2008, 53(1–2): 75–88.
[139] Mar R, Goyal A, Nguyen V, Yang TL, Singhose W. Combined input shaping and feedback control for double-pendulum systems. Mech Syst Signal Proc, 2017, 85: 267–277.
[140] Zhang MH, Ma X, Rong XW, Tian XC, Li YB. Adaptive tracking control for double-pendulum overhead cranes subject to tracking error limitation, parametric uncertainties and external disturbances. Mech Syst Signal Proc, 2016, 76–77: 15–32.
[141] Qian DW, Tong SW, Lee S. Fuzzy-Logic-based control of payloads subjected to double-pendulum motion in overhead cranes. Autom Constr, 2016, 65: 133–143.

2 Crane Dynamics

Abstract: This chapter investigates the double-pendulum-type overhead crane dynamics via Lagrangian mechanics. The double-pendulum-type motions of cranes include two cases such as lumped-mass payload and distributed-mass payload. For each case, the equations of motion are presented by means of the Euler–Lagrange equations. Since the crane models are presented under some ideal assumptions, some uncertainties associated with real applications are discussed next. In addition, the chapter proceeds with the analysis of oscillation frequencies for double-pendulum-type motions based on the methodology of linearization. The analysis distils the essential properties of double-pendulum-type motions.

Keywords: modeling, Lagrangian mechanism, double-pendulum-type motions, linearization

2.1 Introduction

Lagrangian mechanics is a reformulation of classical mechanics introduced by the Italian-French mathematician and astronomer Joseph-Louis Lagrange in 1788 [1]. In Lagrangian mechanics, the trajectory of a system of particles is derived by solving the Lagrange equations in one of two forms, either the Lagrange equations of the first type or the Lagrange equations of the second type. The first type treats constraints explicitly as extra equations, often using Lagrange multipliers, while the second type incorporates the constraints directly by judicious choice of generalized coordinates. In each case, a mathematical function called the Lagrangian is a function of the generalized coordinates, their time derivatives, and time, and contains information about the dynamics of the system.

Compared with Newtonian mechanics, Lagrangian mechanics has no new physics. In Newtonian mechanics, the equations of motion are given by Newton's laws. The second law, net force equals mass times acceleration, is applied to each particle. Instead of forces, Lagrangian mechanics uses the energies in the system. The central quantity of Lagrangian mechanics is the Lagrangian, a function that summarizes the dynamics of the entire system. Overall, the Lagrangian has units of energy, but no single expression for all physical systems. Any function that generates the correct equations of motion, in agreement with physical laws, can be taken as a Lagrangian. It is nevertheless possible to construct general expressions for large classes of applications. Lagrangian mechanics is important not just for its broad applications but also for its role in advancing deep understanding of physics. Lagrangian mechanics is widely used to solve mechanical problems in physics and engineering.

DOI 10.1515/9783110520101-002

By adapting the Lagrangian method, the Lagrangian equation has the form

$$\frac{\mathrm{d}}{\mathrm{d}t}\left(\frac{\partial L}{\partial \dot{q}_i}\right) - \frac{\partial L}{\partial q_i} = T_i \tag{2.1}$$

where L is the Lagrangian, q_i is the generalized coordinate, and T_i is the external force.

In order to illustrate the application of Lagrangian mechanics, a crane system with single-pendulum-type cargo is taken into account as an example. Consider the cargo mass m and its rope length l, which is attached to a trolley with mass M. The trolley can move along a line in the x-direction. The schematic of such a crane is displayed in Figure 2.1. Other symbols in Figure 2.1 are described as the swing angle with respect to the vertical line θ, the trolley position with respect to the origin x, and the driven force applied to the trolley f.

For the purpose of simplification, the following assumptions in Ref. [2] are considered: (1) the payload is regarded as a material particle; (2) the rope is considered as an inflexible rod; (3) compared with the payload mass, the rope mass is ignored; (4) the trolley moves in the x-direction; (5) the payload moves on the x–y surface; and (6) no friction exists in the system.

Consider x and θ as generalized coordinates. The kinetic energy of the crane system [3, 4] can be written as

$$K = \frac{1}{2}M\dot{x}^2 + \frac{1}{2}mv_x^2 + \frac{1}{2}mv_y^2 \tag{2.2}$$

Here $v_x = \dot{x}_m = \dot{x} + l\dot{\theta}\cos\theta$ and $v_y = \dot{y}_m = -l\dot{\theta}\sin\theta$.

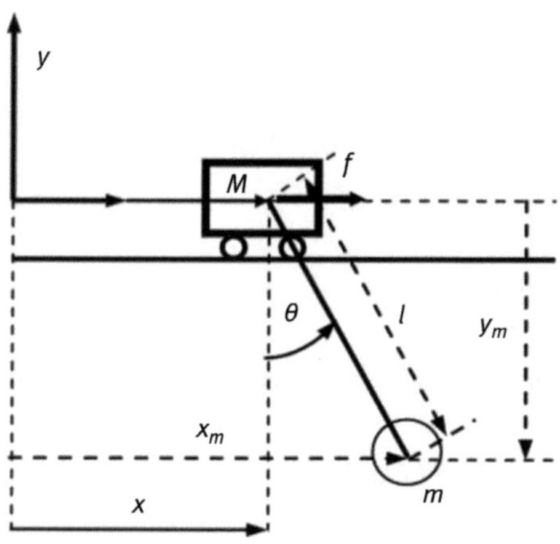

Figure 2.1: Schematic of a single-pendulum-type crane system.

Since the potential energy of the trolley is kept unchanged, the potential energy of the crane in Figure 2.1 is only exhibited by the cargo, formulated as

$$P = mgl(1 - \cos\theta) \tag{2.3}$$

Here g is the gravitational acceleration.

From eqs. (2.2) and (2.3), the Lagrangian of the crane system can be defined as

$$L = K - P = \frac{1}{2}M\dot{x}^2 + \frac{1}{2}mv_x^2 + \frac{1}{2}mv_y^2 - mgl(1 - \cos\theta) \tag{2.4}$$

Differentiating L with respect to x in eq. (2.4) yields

$$\frac{\partial L}{\partial x} = 0 \tag{2.5}$$

Differentiating L with respect to $\dot{x}$ in eq. (2.4) yields

$$\frac{\partial L}{\partial \dot{x}} = M\dot{x} + m(\dot{x} + l\dot{\theta}\cos\theta) \tag{2.6}$$

Further, differentiating eq. (2.6) with respect to time t can have

$$\frac{\mathrm{d}}{\mathrm{d}t}\left(\frac{\partial L}{\partial \dot{x}}\right) = M\ddot{x} + m(\ddot{x} + l\ddot{\theta}\cos\theta - l\dot{\theta}^2\sin\theta) \tag{2.7}$$

Finally, the Lagrangian equation with respect to the generalized coordinate x has the form

$$\frac{\mathrm{d}}{\mathrm{d}t}\left(\frac{\partial L}{\partial \dot{x}}\right) - \frac{\partial L}{\partial x} = (m + M)\ddot{x} + ml(\ddot{\theta}\cos\theta - \dot{\theta}^2\sin\theta) = f \tag{2.8}$$

Take the generalized coordinate θ into account. Differentiating L with respect to θ in eq. (2.4) yields

$$\begin{aligned}\frac{\partial L}{\partial \theta} &= m[(\dot{x} + l\dot{\theta}\cos\theta)(-l\dot{\theta}\sin\theta) + (l\dot{\theta}\sin\theta)(l\dot{\theta}\cos\theta)] - mgl\sin\theta \\ &= -ml\dot{x}\dot{\theta}\sin\theta - mgl\sin\theta\end{aligned} \tag{2.9}$$

Differentiating L with respect to $\dot{\theta}$ in eq. (2.4) yields

$$\frac{\partial L}{\partial \dot{\theta}} = m[(\dot{x} + l\dot{\theta}\cos\theta)(l\cos\theta) + (-l\dot{\theta}\sin\theta)(-l\dot{\theta}\sin\theta)] = ml\dot{x}\cos\theta + ml^2\dot{\theta} \tag{2.10}$$

Further, differentiating eq. (2.10) with respect to time t can have

$$\frac{\mathrm{d}}{\mathrm{d}t}\left(\frac{\partial L}{\partial \dot{\theta}}\right) = ml\ddot{x}\cos\theta - ml\dot{x}\dot{\theta}\sin\theta + ml^2\ddot{\theta} \tag{2.11}$$

Finally, the Lagrangian equation with respect to the generalized coordinate θ has the form

$$\frac{\mathrm{d}}{\mathrm{d}t}\left(\frac{\partial L}{\partial \dot{\theta}}\right) - \frac{\partial L}{\partial \theta} = ml\ddot{x}\cos\theta + ml^2\ddot{\theta} + mgl\sin\theta = 0 \tag{2.12}$$

From eqs. (2.8) and (2.12), the dynamics of this crane system can be obtained by means of the Lagrangian method:

$$\begin{aligned} (m+M)\ddot{x} + ml(\ddot{\theta}\cos\theta - \dot{\theta}^2\sin\theta) &= f \\ ml\ddot{x}\cos\theta + ml^2\ddot{\theta} + mgl\sin\theta &= 0 \end{aligned} \tag{2.13}$$

2.2 Dynamics of Double-Pendulum-Type Crane Systems

This book concentrates on the control problems of double-pendulum-type overhead cranes. Consequently, the dynamics of such kind of crane systems is investigated in light of Lagrangian mechanics.

2.2.1 Two-Dimensional Case

Figure 2.2 illustrates the schematic representation of a two-dimensional double-pendulum-type crane in the Cartesian space. In Figure 2.2, the crane-driven force f can move along the x-direction. The crane system consists of trolley, hook, and payload. In the Cartesian coordinates, the origin is on the left, the trolley position with respect to the origin is described by x (m). In order to define the double-pendulum motions, two angles are defined, that is, the hook angle with respect to the vertical line θ_1 (rad) and the payload angle with respect to the vertical line θ_2 (rad).

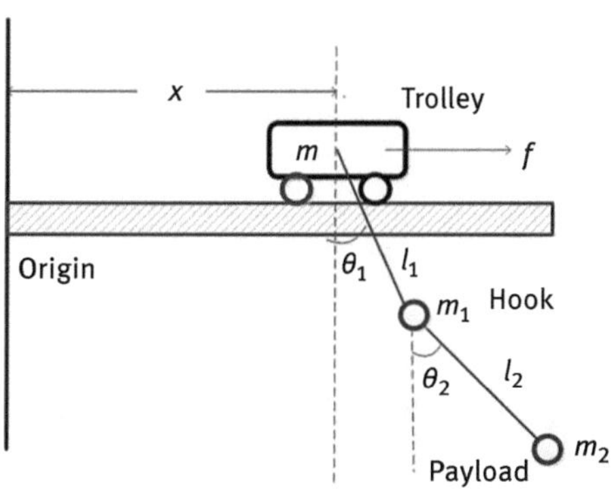

Figure 2.2: Schematic of two-dimensional double-pendulum-type overhead cranes.

Other symbols in Figure 2.2 are trolley mass m_0 (kg), hook mass m_1 (kg), payload mass m_2 (kg), cable length between trolley and hook l_1 (m), and cable length between hook and payload l_2 (m).

Consider some mild and ideal assumptions such as no friction, massless cables, mass-point hook, and mass-point payload. Since the crane system in Figure 2.2 contains three degrees of freedom, three generalized variables are needed to describe its dynamics. In Cartesian space, the trolley position x, the hook angle θ_1, and the payload angle θ_2 are adopted to describe the three degrees of freedom. Take the three variables into account. The following Lagrangian equation [5] can be obtained as

$$\frac{\mathrm{d}}{\mathrm{d}t}\left(\frac{\partial L}{\partial \dot{q}_i}\right) - \frac{\partial L}{\partial q_i} = T_i \tag{2.14}$$

Here $i = 1, 2, 3$; q_i is the generalized coordinate, meaning x, θ_1, and θ_2; T_i is the external force; L indicates the Lagrangian of the crane system.

According to the Lagrangian mechanics, L, the Lagrangian of the crane system can be defined by $L = K - P$, where K means the kinetic energy of the crane system and P denotes the system potential energy. From the aforementioned assumptions, the hook and payload are considered as mass points so that K, the kinetic energy, of the crane system in Figure 2.2 can be written as

$$K = \frac{1}{2}m_0\dot{x}^2 + \frac{1}{2}m_1(v_{1x}^2 + v_{1y}^2) + \frac{1}{2}m_2(v_{2x}^2 + v_{2y}^2) \tag{2.15}$$

here v_{1x} and v_{1y} are the hook velocities in the x-direction and y-direction, respectively, and v_{2x} and v_{2y} are the payload velocities in the x-direction and y-direction, respectively. They can be formulated as

$$v_{1x} = \dot{x} + l_1\dot{\theta}_1 \cos\theta_1 \quad v_{1y} = -l_1\dot{\theta}_1 \sin\theta_1$$
$$v_{2x} = \dot{x} + l_1\dot{\theta}_1 \cos\theta_1 + l_2\dot{\theta}_2 \cos\theta_2 \quad v_{2y} = -l_1\dot{\theta}_1 \sin\theta_1 - l_2\dot{\theta}_2 \sin\theta_2$$

In Figure 2.2, the potential energy of the trolley is kept unchanged. Due to the fact, the potential energy of the crane system is exhibited by the potential energies of the hook and payload determined as

$$P = m_1 g l_1(1 - \cos\theta_1) + m_2 g\left[l_1(1 - \cos\theta_1) + l_2(1 - \cos\theta_2)\right] \tag{2.16}$$

Here g denotes the gravitational acceleration.

Consequently, L, the Lagrangian of the crane system, has the form

$$\begin{aligned} L &= K - P \\ &= \frac{1}{2}m_0\dot{x}^2 + \frac{1}{2}m_1(v_{1x}^2 + v_{1y}^2) + \frac{1}{2}m_2(v_{2x}^2 + v_{2y}^2) - m_1 g l_1(1 - \cos\theta_1) \\ &\quad - m_2 g\left[l_1(1 - \cos\theta_1) + l_2(1 - \cos\theta_2)\right] \end{aligned} \tag{2.17}$$

Take the first generalized coordinate x-into account. Differentiating L in eq. (2.17) with respect to x yields

$$\frac{\partial L}{\partial x} = 0 \tag{2.18}$$

Differentiating L in eq. (2.17) with respect to $\dot{x}$ yields

$$\frac{\partial L}{\partial \dot{x}} = (m_0 + m_1 + m_2)\dot{x} + (m_1 + m_2)l_1\dot{\theta}_1 \cos\theta_1 + m_2 l_2 \dot{\theta}_2 \cos\theta_2 \tag{2.19}$$

Further, differentiating eq. (2.19) with respect to time t can have

$$\begin{aligned}\frac{\mathrm{d}}{\mathrm{d}t}\left(\frac{\partial L}{\partial \dot{x}}\right) &= (m_0 + m_1 + m_2)\ddot{x} + (m_1 + m_2)l_1\ddot{\theta}_1 \cos\theta_1 \\ &\quad - (m_1 + m_2)l_1\dot{\theta}_1^2 \sin\theta_1 + m_2 l_2 \ddot{\theta}_2 \cos\theta_2 - m_2 l_2 \dot{\theta}_2^2 \sin\theta_2\end{aligned} \tag{2.20}$$

Concerning the force f applied to the trolley, the Lagrangian equation with respect to the generalized coordinate x has the form

$$\begin{aligned}\frac{\mathrm{d}}{\mathrm{d}t}\left(\frac{\partial L}{\partial \dot{x}}\right) - \frac{\partial L}{\partial x} &= (m_0 + m_1 + m_2)\ddot{x} + (m_1 + m_2)l_1\ddot{\theta}_1 \cos\theta_1 - (m_1 + m_2)l_1\dot{\theta}_1^2 \sin\theta_1 \\ &\quad + m_2 l_2 \ddot{\theta}_2 \cos\theta_2 - m_2 l_2 \dot{\theta}_2^2 \sin\theta_2 = f\end{aligned} \tag{2.21}$$

Take the second generalized coordinate θ_1 into account. Differentiating L in eq. (2.17) with respect to θ_1 yields

$$\frac{\partial L}{\partial \theta_1} = -(m_1 + m_2)l_1\dot{x}\dot{\theta}_1 \sin\theta_1 - m_2 l_1 l_2 \dot{\theta}_1\dot{\theta}_2 \sin(\theta_1 - \theta_2) - (m_1 + m_2)g l_1 \sin\theta_1 \tag{2.22}$$

Differentiating L in eq. (2.17) with respect to $\dot{\theta}_1$ yields

$$\frac{\partial L}{\partial \dot{\theta}_1} = (m_1 + m_2)(l_1\dot{x}\cos\theta_1 + l_1^2\dot{\theta}_1) + m_2 l_1 l_2 \dot{\theta}_2 \cos(\theta_1 - \theta_2) \tag{2.23}$$

Further, differentiating eq. (2.23) with respect to time t can have

$$\begin{aligned}\frac{\mathrm{d}}{\mathrm{d}t}\left(\frac{\partial L}{\partial \dot{\theta}_1}\right) &= (m_1 + m_2)(l_1\ddot{x}\cos\theta_1 - l_1\dot{x}\dot{\theta}_1 \cos\theta_1 + l_1^2\ddot{\theta}_1) \\ &\quad + m_2 l_1 l_2 \ddot{\theta}_2 \cos(\theta_1 - \theta_2) - m_2 l_1 l_2 \dot{\theta}_1\dot{\theta}_2 \sin(\theta_1 - \theta_2) + m_2 l_1 l_2 \dot{\theta}_2^2 \sin(\theta_1 - \theta_2)\end{aligned} \tag{2.24}$$

Concerning no inputs applied to the hook, the Lagrangian equation with respect to the generalized coordinate θ_1 has the form

$$\begin{aligned}\frac{\mathrm{d}}{\mathrm{d}t}\left(\frac{\partial L}{\partial \dot{\theta}_1}\right)-\frac{\partial L}{\partial \theta_1}&=(m_1+m_2)l_1\ddot{x}\cos\theta_1+(m_1+m_2)l_1^2\ddot{\theta}_1\\&\quad+m_2l_1l_2\ddot{\theta}_2\cos(\theta_1-\theta_2)+m_2l_1l_2\dot{\theta}_2^2\sin(\theta_1-\theta_2)\\&\quad+(m_1+m_2)gl_1\sin\theta_1=0\end{aligned}\tag{2.25}$$

Take the third second generalized coordinate θ_2 into account. Differentiating L in eq. (2.17) with respect to θ_2 yields

$$\frac{\partial L}{\partial \theta_2}=-m_2l_1\dot{x}\dot{\theta}_2\sin\theta_2+m_2l_1l_2\dot{\theta}_1\dot{\theta}_2\sin(\theta_1-\theta_2)-m_2gl_2\sin\theta_2\tag{2.26}$$

Differentiating L in eq. (2.17) with respect to $\dot{\theta}_2$ yields

$$\frac{\partial L}{\partial \dot{\theta}_2}=m_2l_1\dot{x}\cos\theta_1+m_2l_2^2\dot{\theta}_2+m_2l_1l_2\dot{\theta}_1\cos(\theta_1-\theta_2)\tag{2.27}$$

Further, differentiating eq. (2.27) with respect to time t can have

$$\begin{aligned}\frac{\mathrm{d}}{\mathrm{d}t}\left(\frac{\partial L}{\partial \dot{\theta}_2}\right)&=m_2l_2\ddot{x}\cos\theta_2-m_2l_2\dot{x}\dot{\theta}_2\cos\theta_2+m_2l_2^2\ddot{\theta}_2+m_2l_1l_2\ddot{\theta}_1\cos(\theta_1-\theta_2)\\&\quad-m_2l_1l_2\dot{\theta}_1^2\sin(\theta_1-\theta_2)+m_2l_1l_2\dot{\theta}_1\dot{\theta}_2\sin(\theta_1-\theta_2)\end{aligned}\tag{2.28}$$

Concerning no inputs applied to the payload, the Lagrangian equation with respect to the generalized coordinate θ_2 has the form

$$\begin{aligned}\frac{\mathrm{d}}{\mathrm{d}t}\left(\frac{\partial L}{\partial \dot{\theta}_2}\right)-\frac{\partial L}{\partial \theta_2}&=m_2l_2\ddot{x}\cos\theta_2+m_2l_2^2\ddot{\theta}_2+m_2l_1l_2\ddot{\theta}_1\cos(\theta_1-\theta_2)\\&\quad-m_2l_1l_2\dot{\theta}_1^2\sin(\theta_1-\theta_2)+m_2gl_2\sin\theta_2=0\end{aligned}\tag{2.29}$$

From eqs. (2.21), (2.25), and (2.29), the dynamic model of the crane system can be written as

$$\begin{aligned}&(m_0+m_1+m_2)\ddot{x}+(m_1+m_2)l_1\ddot{\theta}_1\cos\theta_1+m_2l_2\ddot{\theta}_2\cos\theta_2\\&\qquad-(m_1+m_2)l_1\dot{\theta}_1^2\sin\theta_1-m_2l_2\dot{\theta}_2^2\sin\theta_2=f\\&(m_1+m_2)l_1\ddot{x}\cos\theta_1+(m_1+m_2)l_1^2\ddot{\theta}_1+m_2l_1l_2\ddot{\theta}_2\cos(\theta_1-\theta_2)\\&\qquad+m_2l_1l_2\dot{\theta}_2^2\sin(\theta_1-\theta_2)+(m_1+m_2)gl_1\sin\theta_1=0\\&m_2l_2\ddot{x}\cos\theta_2+m_2l_1l_2\ddot{\theta}_1\cos(\theta_1-\theta_2)+m_2l_2^2\ddot{\theta}_2-m_2l_1l_2\dot{\theta}_1^2\sin(\theta_1-\theta_2)\\&\qquad+m_2gl_2\sin\theta_2=0\end{aligned}\tag{2.30}$$

Rearrange eq. (2.30) in the form of matrix. The dynamics [6] can be rewritten as

$$\mathbf{M}(\mathbf{q})\,\ddot{\mathbf{q}} + \mathbf{C}(\mathbf{q},\dot{\mathbf{q}})\,\dot{\mathbf{q}} + \mathbf{G}(\mathbf{q}) = \tau \tag{2.31}$$

Here $\mathbf{q} = [x, \theta_1, \theta_2]^T$ is a vector of the three generalized coordinates, $\tau = [f, 0, 0]^T$ is a vector of the generalized force, g is the gravitational acceleration, $\mathbf{M}(\mathbf{q})$ is a 3×3 inertia matrix, $\mathbf{C}(\mathbf{q},\dot{\mathbf{q}})\,\dot{\mathbf{q}}$ is a vector of Coriolis and centripetal torques, and $\mathbf{G}(\mathbf{q})$ is a vector of the gravitational term. $\mathbf{M}(\mathbf{q})$, $\mathbf{C}(\mathbf{q},\dot{\mathbf{q}})$, and $\mathbf{G}(\mathbf{q})$ are given as

$$\mathbf{M}(\mathbf{q}) = \begin{bmatrix} m_0 + m_1 + m_2 & (m_1 + m_2)\,l_1 \cos\theta_1 & m_2 l_2 \cos\theta_2 \\ (m_1 + m_2)\,l_1 \cos\theta_1 & (m_1 + m_2)\,l_1^2 & m_2 l_1 l_2 \cos(\theta_1 - \theta_2) \\ m_2 l_2 \cos\theta_2 & m_2 l_1 l_2 \cos(\theta_1 - \theta_2) & m_2 l_2^2 \end{bmatrix}$$

$$\mathbf{C}(\mathbf{q},\dot{\mathbf{q}}) = \begin{bmatrix} 0 & -(m_1 + m_2) l_1 \dot{\theta}_1 \sin\theta_1 & -m_2 l_2 \dot{\theta}_2 \sin\theta_2 \\ 0 & 0 & m_2 l_1 l_2 \dot{\theta}_2 \sin(\theta_1 - \theta_2) \\ 0 & -m_2 l_1 l_2 \dot{\theta}_1 \sin(\theta_1 - \theta_2) & 0 \end{bmatrix}$$

$$\mathbf{G}(\mathbf{q}) = \begin{bmatrix} 0 & (m_1 + m_2)\,g l_1 \sin\theta_1 & m_2 g l_2 \sin\theta_2 \end{bmatrix}^T$$

According to the concept of state space, eq. (2.31) can be written by a set of first-order ordinary differential equations:

$$\begin{aligned} \dot{x}_1 &= x_2 \\ \dot{x}_2 &= f_1(\mathbf{x}) + b_1(\mathbf{x})u \\ \dot{x}_3 &= x_4 \\ \dot{x}_4 &= f_2(\mathbf{x}) + b_2(\mathbf{x})u \\ \dot{x}_5 &= x_6 \\ \dot{x}_6 &= f_3(\mathbf{x}) + b_3(\mathbf{x})u \end{aligned} \tag{2.32}$$

Here $\mathbf{x}$ is defined by $[x_1, x_2, x_3, x_4, x_5, x_6]^T$; $x_1 = x$; $x_3 = \theta_1$; $x_5 = \theta_2$; x_2 is the trolley velocity; x_4 is the angular velocity of the hook; x_6 is the angular velocity of the payload; $u = f$ denotes the control input; $f_i(\mathbf{x})$ and $b_i(\mathbf{x})$ $(i = 1, 2, 3)$ are nonlinear functions of the vector $\mathbf{x}$, formulated by $f_i(\mathbf{x}) = \Gamma_i/\Delta$ and $b_i(\mathbf{x}) = \mathrm{T}_i/\Delta$. Here Γ_i, T_i, and Δ are determined as

$$\begin{aligned} \Delta = {} & (m_1 + m_2) m_2 l_1^2 l_2^2 [(m_0 + m_1 + m_2) - (m_1 + m_2)\cos^2(x_3)] - m_2^2 l_1^2 l_2^2 [(m_1 + m_2)\cos^2(x_5) \\ & + (m_0 + m_1 + m_2)\cos^2(x_3 - x_5) - 2(m_1 + m_2)\cos(x_3)\cos(x_5)\cos(x_3 - x_5)] \end{aligned}$$

$$\begin{aligned} \Gamma_1 = {} & [(m_1 + m_2) m_2 l_1^2 l_2^2 - m_2^2 l_1^2 l_2^2 \cos^2(x_3 - x_5)][(m_1 + m_2) l_1 x_4^2 \sin(x_3) + m_2 l_2 x_6^2 \sin(x_5)] \\ & + [(m_1 + m_2) m_2 l_1 l_2^2 \cos(x_3) - m_2^2 l_1 l_2^2 \cos(x_5)\cos(x_3 - x_5)][m_2 l_1 l_2 x_6^2 \sin(x_3 - x_5) \\ & + (m_1 + m_2) g l_1 \sin(x_3)] + [(m_1 + m_2) m_2 l_1^2 l_2 \cos(x_5) - m_2 l_1^2 l_2 \cos(x_3)\cos(x_3 - x_5)] \\ & [-m_2 l_1 l_2 x_4^2 \sin(x_3 - x_5) + m_2 g l_2 \sin(x_5)] \end{aligned}$$

$$T_1 = (m_1 + m_2)m_2 l_1^2 l_2^2 - m_2^2 l_1^2 l_2^2 \cos^2(x_3 - x_5)$$

$$\begin{aligned}\Gamma_2 = {} & [m_2^2 l_1 l_2^2 \cos(x_5)\cos(x_3 - x_5) - (m_1 + m_2)m_2 l_1 l_2^2 \cos(x_3)][(m_1 + m_2)l_1 x_4^2 \sin(x_3) \\ & + m_2 l_2 x_6^2 \sin(x_5)] + [m_2^2 l_2^2 \cos^2(x_5) - (m_0 + m_1 + m_2)m_2 l_2^2][m_2 l_1 l_2 x_6^2 \sin(x_3 - x_5) \\ & + (m_1 + m_2)g l_1 \sin(x_3)] + [(m_0 + m_1 + m_2)m_2 l_1 l_2 \cos(x_3 - x_5) \\ & - (m_1 + m_2)m_2 l_1 l_2 \cos(x_3)\cos(x_5)][-m_2 l_1 l_2 x_4^2 \sin(x_3 - x_5) + m_2 g l_2 \sin(x_5)]\end{aligned}$$

$$T_2 = m_2^2 l_1^2 l_2^2 \cos(x_5)\cos(x_3 - x_5) - (m_1 + m_2)m_2 l_1 l_2^2 \cos(x_3)$$

$$\begin{aligned}\Gamma_3 = {} & \{(m_1 + m_2)m_2 l_1^2 l_2[\cos(x_3)\cos(x_3 - x_5) - \cos(x_5)]\}[(m_1 + m_2)l_1 x_4^2 \sin(x_3) \\ & + m_2 l_2 x_6^2 \sin(x_5)] + [(m_0 + m_1 + m_2)m_2 l_1 l_2 \cos(x_3 - x_5) \\ & - (m_1 + m_2)m_2 l_1 l_2 \cos(x_3)\cos(x_5)][m_2 l_1 l_2 x_6^2 \sin(x_3 - x_5) + (m_1 + m_2)g l_1 \sin(x_3)] \\ & + [(m_1 + m_2)^2 l_1^2 \cos^2(x_3) - (m_0 + m_1 + m_2)(m_1 + m_2)l_1^2] \\ & \times [-m_2 l_1 l_2 x_4^2 \sin(x_3 - x_5) + m_2 g l_2 \sin(x_5)]\end{aligned}$$

$$T_3 = (m_1 + m_2)m_2 l_1^2 l_2 \cos(x_3)\cos(x_3 - x_5) - (m_1 + m_2)m_2 l_1^2 l_2 \cos(x_5)$$

2.2.2 Three-Dimensional Case

The crane system in Figure 2.2 is two-dimensional and it is assumed to be with mass-point hook and mass-point payload. Take the same assumptions into account. Dynamics (2.30), its matrix form (2.31), and its state equations form (2.32) can be extended to three-dimensional case. Such an extension increases the system complexity but does not challenge.

In some operating conditions, a crane system may exhibit double-pendulum dynamics with distributed-mass payload. In Figure 2.3, the payload becomes a bar that can hardly be treated as a point-mass one. The crane system not only increases the complexity from the perspective of dimension but also the double-pendulum motions with distributed-mass payload involve rigid body dynamics. Under this condition, modeling such a crane system becomes challenging.

In the $X - Y$ plane, the trolley in Figure 2.3 can simultaneously move in both x- and y-directions. The reference system with coordinates $X_1 - Y_1 - Z_1$ is attached to the trolley with Z_1 vertically downward. The hook is connected to the trolley by a light, flexible cable, with the cable mass considered negligible in comparison with that of the hook and the payload. The hook is assumed to be mass point and its mass is m_1. The cable length between the hook and the trolley is taken as a constant l_1. In three-dimensional space, the hook angle is spatial, measured by two variables, θ_{1x} and θ_{1y}. The two variables are the projections of the cable onto the $X - Z$ and $Y - Z$ planes, respectively.

In Figure 2.3, the second reference frame with coordinates $X_2 - Y_2 - Z_2$, parallel to the first reference system, has its origin at the hook. The second reference system is

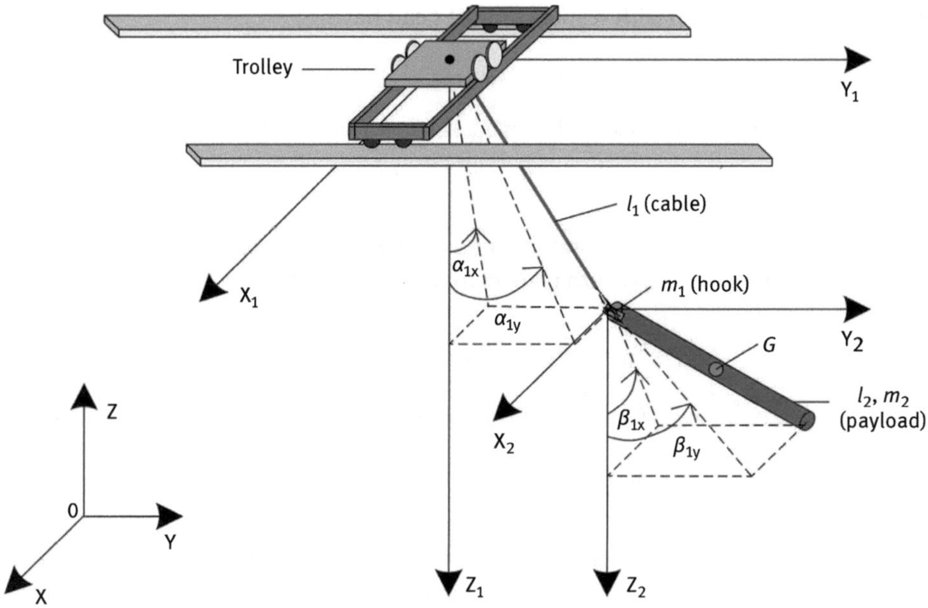

Figure 2.3: Schematic of a three-dimensional crane system with distributed-mass payload.

utilized to describe the swing rotation of the payload. Similarly, the payload rotation can be described by two angles, θ_{2x} and θ_{2y}, which are projections of the payload angle onto the $X - Z$ and $Y - Z$ planes, respectively.

Other symbols in Figure 2.3 are payload mass m_2, payload length l_2, and the payload center of gravity G. From Figure 2.3, the instantaneous location of hook $[x_{\mathrm{H}}, y_{\mathrm{H}}, z_{\mathrm{H}}]^T$ with respect to an inertial coordinate system can have the form

$$\begin{aligned} x_{\mathrm{H}} &= x_{\mathrm{T}} + l_1 \sin\theta_{1x} \\ y_{\mathrm{H}} &= y_{\mathrm{T}} + l_1 \cos\theta_{1x} \sin\theta_{1y} \\ z_{\mathrm{H}} &= l_1 \cos\theta_{1x} \cos\theta_{1y} \end{aligned} \tag{2.33}$$

Here x_T and y_T are the trolley positions in the inertial reference frame.

The instantaneous location of the payload center of gravity $[x_{\mathrm{G}}, y_{\mathrm{G}}, z_{\mathrm{G}}]^T$ with respect to an inertial coordinate system can have the form

$$\begin{aligned} x_{\mathrm{G}} &= x_{\mathrm{T}} + l_1 \sin\theta_{1x} + \frac{l_2}{2}\sin\theta_{2x} \\ y_{\mathrm{G}} &= y_{\mathrm{T}} + l_1 \cos\theta_{1x} \sin\theta_{1y} + \frac{l_2}{2}\cos\theta_{2x}\sin\theta_{2y} \\ z_{\mathrm{G}} &= l_1 \cos\theta_{1x} \cos\theta_{1y} + \frac{l_2}{2}\cos\theta_{2x}\cos\theta_{2y} \end{aligned} \tag{2.34}$$

Differentiating eq. (2.33) with respect to time t can have

$$\vec{v}_{\mathrm{H}}:\quad \begin{aligned} \dot{x}_{\mathrm{H}} &= \dot{x}_{\mathrm{T}} + l_1\dot{\theta}_{1x}\cos\theta_{1x} \\ \dot{y}_{\mathrm{H}} &= \dot{y}_{\mathrm{T}} - l_1\dot{\theta}_{1x}\sin\theta_{1x}\sin\theta_{1y} + l_1\dot{\theta}_{1y}\cos\theta_{1x}\cos\theta_{1y} \\ \dot{z}_{\mathrm{H}} &= -l_1\dot{\theta}_{1x}\sin\theta_{1x}\cos\theta_{1y} - l_1\dot{\theta}_{1y}\cos\theta_{1x}\sin\theta_{1y} \end{aligned} \tag{2.35}$$

where $\vec{v}_{\mathrm{H}} = [\dot{x}_{\mathrm{H}}\ \dot{y}_{\mathrm{H}}\ \dot{z}_{\mathrm{H}}]^T$ is the instantaneous velocity of the hook.

Differentiating eq. (2.34) with respect to time t can have

$$\vec{v}_{\mathrm{G}}:\quad \begin{aligned} \dot{x}_{\mathrm{G}} &= \dot{x}_{\mathrm{H}} + \frac{l_2}{2}\dot{\theta}_{2x}\cos\theta_{2x} \\ \dot{y}_{\mathrm{G}} &= \dot{y}_{\mathrm{H}} - \frac{l_2}{2}\dot{\theta}_{2x}\sin\theta_{2x}\sin\theta_{2y} + \frac{l_2}{2}\dot{\theta}_{2y}\cos\theta_{2x}\cos\theta_{2y} \\ \dot{z}_{\mathrm{G}} &= \dot{z}_{\mathrm{H}} - \frac{l_2}{2}\dot{\theta}_{2x}\sin\theta_{2x}\cos\theta_{2y} - \frac{l_2}{2}\dot{\theta}_{2y}\cos\theta_{2x}\sin\theta_{2y} \end{aligned} \tag{2.36}$$

where $\vec{v}_{\mathrm{G}} = [\dot{x}_{\mathrm{G}}\ \dot{y}_{\mathrm{G}}\ \dot{z}_{\mathrm{G}}]^T$ is the instantaneous velocity of the payload center of gravity.

Take the four variables θ_{1x}, θ_{1y}, θ_{2x}, and θ_{2y} into account as generalized coordinates. The following Lagrangian equation can be obtained to formulate the double-pendulum dynamics with distributed-mass payload:

$$\frac{\mathrm{d}}{\mathrm{d}t}\left(\frac{\partial L}{\partial \dot{q}_i}\right) - \left(\frac{\partial L}{\partial q_i}\right) = 0 \tag{2.37}$$

Here $i = 1, 2, 3$, and 4; q_i is the generalized coordinate, meaning θ_{1x}, θ_{1y}, θ_{2x}, and θ_{2y}, respectively. The right-hand side in eq. (2.37) is set to be zeroth because there are no external inputs applied to the hook as well as the payload. L is the Lagrangian of the double-pendulum system, defined as $L = K - P$, where K means the kinetic energy of the double-pendulum system and P denotes the potential energy of the double-pendulum system. In Figure 2.3, the potential energies of the double-pendulum motions are determined as

$$P = m_1 g l_1(1 - \cos\theta_{1x}\cos\theta_{1y}) + m_2 g[l_1(1 - \cos\theta_{1x}\cos\theta_{1y}) + \frac{l_2}{2}(1 - \cos\theta_{2x}\cos\theta_{2y})] \tag{2.38}$$

On the other hand, the kinetic energy K of the double-pendulum system is complicated by the existence of the distributed-mass payload. To quantify the rotational kinetic energy of the payload, the angular velocity of the payload ϖ_{G} at its center of gravity and the moment of inertia tensor matrix I_{G} in a common coordinate system are required. The load inertia tensor can be expressed by using a coordinate system $X_{2c} - Y_{2c} - Z_{2c}$ fixed to the payload and along the principal axes, formulated as

$$I_{\mathrm{G}} = \begin{bmatrix} I_{xx} & 0 & 0 \\ 0 & I_{yy} & 0 \\ 0 & 0 & I_{zz} \end{bmatrix}_{X_{2c} - Y_{2c} - Z_{2c}} \tag{2.39}$$

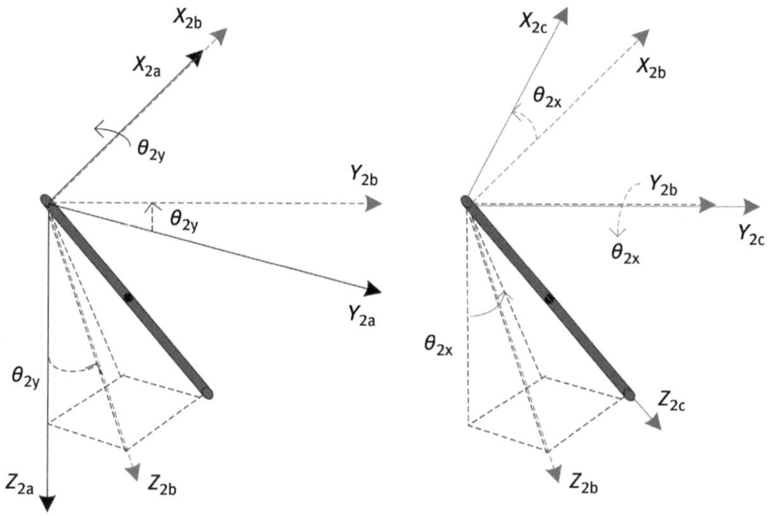

Figure 2.4: Schematic of successive rotations to reach the body coordinates of the payload.

Similarly, the angular velocity ϖ_G should also be expressed in the same coordinate system $X_{2c} - Y_{2c} - Z_{2c}$. As shown in Figure 2.4, the coordinate system $X_{2a} - Y_{2a} - Z_{2a}$ moves through two rotations about the main axes to become aligned to the load body axis $X_{2c} - Y_{2c} - Z_{2c}$.

The angular velocity ϖ_G in Figure 2.4 can be calculated as

$$\varpi_G = -\dot{\theta}_{2y}\vec{i}_{2b} + \dot{\theta}_{2x}\vec{j}_{2c} \tag{2.40}$$

The transformation from the coordinate system $X_{2b} - Y_{2b} - Z_{2b}$ to the coordinate system $X_{2c} - Y_{2c} - Z_{2c}$ can be determined as

$$T_{(2b\to 2c)} = \begin{bmatrix} \cos\theta_{2x} & 0 & -\sin\theta_{2x} \\ 0 & 1 & 0 \\ \sin\theta_{2x} & 0 & \cos\theta_{2x} \end{bmatrix} \tag{2.41}$$

Therefore, the angular velocity ϖ_G in the coordinate system $X_{2c} - Y_{2c} - Z_{2c}$ can have the form

$$\varpi_G = \begin{bmatrix} -\dot{\theta}_{2y}\cos\theta_{2x} \\ \dot{\theta}_{2x} \\ -\dot{\theta}_{2y}\sin\theta_{2x} \end{bmatrix}_{X_{2c}-Y_{2c}-Z_{2c}} \tag{2.42}$$

Totally, the kinetic energy of the crane system K in Figure 2.2 can be determined as

$$K = \frac{1}{2}m_1\vec{v}_H^T\vec{v}_H + \frac{1}{2}m_2\vec{v}_G^T\vec{v}_G + \frac{1}{2}\varpi_G^T[I_G]\varpi_G \tag{2.43}$$

Substituting eqs. (2.38) and (2.43) into eq. (2.37) and differentiating the Lagrangian L with respect to the generalized coordinates yields the double-pendulum dynamics with distributed-mass payload in Figure 2.3 [7]:

$$\begin{aligned}(m_1 l_1^2 + m_2 l_1^2)\ddot{\theta}_{1x} &+ (m_1+m_2)l_1\ddot{x}_T\cos\theta_{1x} - (m_1+m_2)l_1\ddot{y}_T\sin\theta_{1x}\sin\theta_{1y}\\ &+ (m_1+m_2)l_1^2(\dot{\theta}_{1y})^2\cos\theta_{1x}\sin\theta_{1x} + \frac{1}{2}m_2l_1l_2\ddot{\theta}_{2x}\cos\theta_{1x}\cos\theta_{2x}\\ &- \frac{1}{2}m_2l_1l_2(\dot{\theta}_{2x})^2\cos\theta_{1x}\sin\theta_{2x} + \frac{1}{2}m_2l_1l_2\ddot{\theta}_{2x}\sin\theta_{1x}\sin\theta_{2x}\cos(\theta_{1y}-\theta_{2y})\\ &+ \frac{1}{2}m_2l_1l_2(\dot{\theta}_{2x})^2 + (\dot{\theta}_{2y})^2\sin\theta_{1x}\cos\theta_{2x}\cos(\theta_{1y}-\theta_{2y})\\ &+ \frac{1}{2}m_2l_1l_2\ddot{\theta}_{2y}\sin\theta_{1x}\cos\theta_{2x}\sin(\theta_{2y}-\theta_{1y}) + m_2l_1l_2\dot{\theta}_{2x}\dot{\theta}_{2y}\sin\theta_{1x}\sin\theta_{2x}\\ &\sin(\theta_{1y}-\theta_{2y}) + (m_1+m_2)gl_1\sin\theta_{1x}\cos\theta_{1y} = 0 \end{aligned} \tag{2.44}$$

$$\begin{aligned}(m_1+m_2)l_1^2\cos^2\theta_{1x}\ddot{\theta}_{1y} &- 2(m_1+m_2)l_1^2\dot{\theta}_{1x}\dot{\theta}_{1y}\cos\theta_{1x}\sin\theta_{1x} + (m_1+m_2)l_1\ddot{y}_T\cos\theta_{1x}\cos\theta_{1y}\\ &+ \frac{1}{2}m_2l_1l_2\ddot{\theta}_{2y}\cos\theta_{1x}\cos\theta_{2x}\cos(\theta_{1y}-\theta_{2y}) + \frac{1}{2}m_2l_1l_2(\dot{\theta}_{2x})^2\\ &+ (\dot{\theta}_{2y})^2\cos\theta_{1x}\cos\theta_{2x}\sin(\theta_{1y}-\theta_{2y}) + \frac{1}{2}m_2l_1l_2\ddot{\theta}_{2x}\cos\theta_{1x}\sin\theta_{2x}\sin(\theta_{1y}-\theta_{2y})\\ &- m_2l_1l_2\dot{\theta}_{2x}\dot{\theta}_{2y}\cos\theta_{1x}\sin\theta_{2x}\cos(\theta_{1y}-\theta_{2y}) + (m_1+m_2)gl_1\cos\theta_{1x}\sin\theta_{1y} = 0\end{aligned} \tag{2.45}$$

$$\begin{aligned}\left(m_2\frac{l_2^2}{4} + I_{yy}\right)\ddot{\theta}_{2x} &+ \frac{1}{2}m_2l_2\ddot{x}_T\cos\theta_{2x} - \frac{1}{2}m_2l_2\ddot{y}_T\sin\theta_{2x}\sin\theta_{2y} + \frac{1}{2}m_2l_1l_2\ddot{\theta}_{1x}\cos\theta_{1x}\cos\theta_{2x}\\ &- \frac{1}{2}m_2l_1l_2(\dot{\theta}_{1x})^2\sin\theta_{1x}\cos\theta_{2x} + \frac{1}{2}m_2l_1l_2\ddot{\theta}_{1x}\sin\theta_{1x}\sin\theta_{2x}\cos(\theta_{1y}-\theta_{2y})\\ &+ \frac{1}{2}m_2l_1l_2(\dot{\theta}_{1x})^2 + (\dot{\theta}_{1y})^2)\cos\theta_{1x}\sin\theta_{2x}\cos(\theta_{1y}-\theta_{2y})\\ &- m_2l_1l_2\dot{\theta}_{1x}\dot{\theta}_{1y}\sin\theta_{1x}\sin\theta_{2x}\sin(\theta_{1y}-\theta_{2y}) + m_2\frac{l_2^2}{4}(\dot{\theta}_{2y})^2\cos\theta_{2x}\sin\theta_{2x}\\ &+ (I_{xx}-I_{zz})(\dot{\theta}_{2y})^2\cos\theta_{2x}\sin\theta_{2x} + m_2g\frac{l_2}{2}\cos\theta_{2x}\sin\theta_{2y} = 0\end{aligned} \tag{2.46}$$

$$\begin{aligned}\left(m_2\frac{l_2^2}{4}\cos^2\theta_{2x} + I_{xx}\cos^2\theta_{2x} + I_{zz}\sin^2\theta_{2x}\right)\ddot{\theta}_{2y} &- m_2\frac{l_2^2}{2}\dot{\theta}_{2x}\dot{\theta}_{2y}\cos\theta_{2x}\sin\theta_{2x}\\ &+ \frac{1}{2}m_2l_2\ddot{y}_T\cos\theta_{2x}\cos\theta_{2y} + \frac{1}{2}m_2l_1l_2\ddot{\theta}_{1x}\sin\theta_{1x}\cos\theta_{2x}\sin(\theta_{2y}-\theta_{1y})\\ &+ \frac{1}{2}m_2l_1l_2(\dot{\theta}_{1x}^2 + \dot{\theta}_{1y}^2)\cos\theta_{1x}\cos\theta_{2x}\sin(\theta_{2y}-\theta_{1y})\\ &+ \frac{1}{2}m_2l_1l_2\ddot{\theta}_{1y}\cos\theta_{1x}\cos\theta_{2x}\cos(\theta_{1y}-\theta_{2y}) - m_2l_1l_2\dot{\theta}_{1x}\dot{\theta}_{1y}\sin\theta_{1x}\cos\theta_{2x}\cos(\theta_{2y}-\theta_{1y})\\ &+ (I_{xx}-I_{zz})\dot{\theta}_{2x}\dot{\theta}_{2y}\sin(2\theta_{2x}) + m_2g\frac{l_2}{2}\cos\theta_{2x}\sin\theta_{2y} = 0\end{aligned} \tag{2.47}$$

From eq. (2.44) to eq. (2.47), the four highly coupled equations of motion capture the double-pendulum dynamics of the crane system with distributed-mass payload. From the perspective of simplification, only the double-pendulum dynamics are taken into consideration. The trolley dynamics are not considered here because the trolley is just translational and its dynamics can be extended from the two-dimensional case.

Our motivation of modeling the crane system with distributed-mass payload is to reveal its complexity. In the rest of the book, the two-dimensional double-pendulum-type crane system with mass-point payload in Figure 2.2 is taken as a benchmark to illustrate its control design methods.

2.3 Insight into Double-Pendulum Motions

In mathematics, linearization refers to finding the linear approximation to a function at a given point. Inherently, the linearization of a function is the first-order term of its Taylor expansion around the point of interest. In the study of dynamical systems, linearization is a powerful tool to assess the local stability of an equilibrium point of a system of nonlinear differential equations or discrete dynamical systems.

Linearization makes it possible to use developed tools for studying linear systems to analyze the behavior of a nonlinear function near an equilibrium point. The nonlinear double-pendulum-type overhead crane models in various forms are shown in eqs. (2.30), (2.31), and (2.32). In mathematics, these models are equal to each other.

2.3.1 Linearization

In order to illustrate the linearization of the crane system in Figure 2.2, model (2.31) in the form of matrix is taken into consideration. In the model, $\theta_1 = \theta_2 = 0$ is the sole stable equilibrium point so that eq. (2.31) can be linearized around $\theta_1 = 0$ and $\theta_2 = 0$. The linearized crane model [8, 9] can be written as

$$\bar{M}\ddot{\mathbf{q}} + K\mathbf{q} = 0 \tag{2.48}$$

In the linearized model (2.48), the matrices $\bar{M}$ and K are determined as

$$\bar{M} = \begin{bmatrix} m_0 + m_1 + m_2 & (m_1 + m_2)l_1 & m_2 l_2 \\ (m_1 + m_2)l_1 & (m_1 + m_2)l_1^2 & m_2 l_1 l_2 \\ m_2 l_2 & m_2 l_1 l_2 & m_2 l_2^2 \end{bmatrix}$$

$$K = \begin{bmatrix} 0 & 0 & 0 \\ 0 & (m_1 + m_2)gl_1 & 0 \\ 0 & 0 & m_2 g l_2 \end{bmatrix}$$

The two natural frequencies of the double-pendulum-type overhead crane system can be obtained by the nonzero eigenvalues of the matrix $-\bar{M}K$. Their expressions can be formulated as

$$\omega_1 = \sqrt{\frac{g}{2}(\alpha - \sqrt{\beta})} \tag{2.49}$$

$$\omega_2 = \sqrt{\frac{g}{2}(\alpha + \sqrt{\beta})} \tag{2.50}$$

Here the parameter α is given as

$$\alpha = \frac{m_1 + m_2}{m_1}\left(\frac{1}{l_1} + \frac{1}{l_2}\right) \tag{2.51}$$

Other parameter β is formulated as

$$\beta = \left(\frac{m_1 + m_2}{m_1}\right)^2 \left(\frac{1}{l_1} + \frac{1}{l_2}\right)^2 - 4\left(\frac{m_1 + m_2}{m_1}\right)\frac{1}{l_1 l_2} \tag{2.52}$$

The results seem to indicate that an oscillation control scheme would need more robustness to variations in the second mode than in the first mode. However, if the amplitude of the second mode is very small compared to the first mode, then the controller does not need to address the second mode. The relative contribution of the two modes can be examined by breaking the overall dynamic response into components arising from ω_1 and ω_2.

The responses of the two swing angles, θ_1 and θ_2, to an impulse of magnitude A, introduced at time t_0, are

$$\theta_1 = \frac{A\omega_1(1 + \omega_2^2 l_1 \gamma)}{k}\sin[\omega_1(t - t_0)] - \frac{A\omega_2(1 + \omega_1^2 l_1 \gamma)}{k}\sin[\omega_2(t - t_0)] \tag{2.53}$$

$$\theta_2 = \frac{A\omega_1}{k}\sin[\omega_1(t - t_0)] - \frac{A\omega_2}{k}\sin[\omega_2(t - t_0)] \tag{2.54}$$

Here

$$\gamma = -\frac{g(1 + R)}{\omega_1^2 \omega_2^2 l_1 l_2} \text{ and } k = \sqrt{\beta} l_1 g$$

If we assume small angles, then the impulse response of the double-pendulum-type payload in the horizontal direction can be approximated as

$$x = C_1 \sin(\omega_1 t + \psi_1) + C_2 \sin(\omega_2 t + \psi_2) \tag{2.55}$$

Here

$$C_1 = \frac{\omega_1 l_1[1+\omega_2^2\gamma(l_1+l_2)]}{k}\sqrt{\left(\sum_{j=1}^{n} A_j\cos(\omega_1 t_j)\right)^2 + \left(\sum_{j=1}^{n} A_j\sin(\omega_1 t_j)\right)^2} \tag{2.56}$$

$$C_2 = -\frac{\omega_2 l_1[1+\omega_1^2\gamma(l_1+l_2)]}{k}\sqrt{\left(\sum_{j=1}^{n} A_j\cos(\omega_2 t_j)\right)^2 + \left(\sum_{j=1}^{n} A_j\sin(\omega_2 t_j)\right)^2} \tag{2.57}$$

The two coefficients in eqs. (2.56) and (2.57), C_1 and C_2, indicate the contributions of each mode to the overall payload response.

2.3.2 Model Analysis via Matlab

From eqs. (2.49) and (2.50), the two natural frequencies not only depend on the length of the cables but also depend on the masses of payload and hook [10]. It is interesting to investigate how the frequencies change as a function of the system physical parameters. To simplify this problem, $R = m_2/m_1$ is defined as the payload-to-hook mass ratio, the cable length between hook and payload l_2 is considered as a variable when the total length l determined by l_1 plus l_2 is held constant at 6 m.

In Figure 2.5, ω_1 changes very little for a constant $l_1 + l_2$. It corresponds closely to the frequency of a single pendulum with the length of $l_1 + l_2$. On the other hand, the value of ω_1 is maximized for a constant $l_1 + l_2$ when the two cable lengths are equal to $\frac{l_1+l_2}{2}$ but it can be dramatically changed by the hoisting operation.

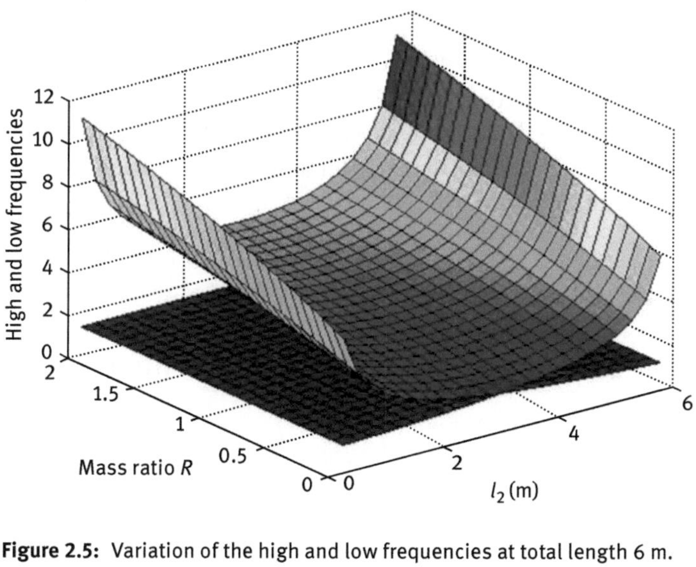

Figure 2.5: Variation of the high and low frequencies at total length 6 m.

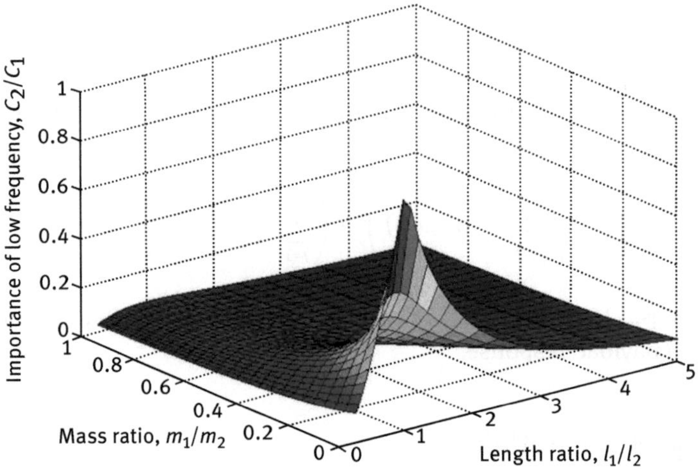

Figure 2.6: Ratio of high-frequency amplitude to low-frequency amplitude.

Furthermore, the high frequency ω_2 has a strong dependence on the cable length l_2, the value of ω_2 varies substantially more than the value of ω_1, the contribution of ω_2 to the problematic swing amplitude is particularly large for a constant $l_1 + l_2$ when the two cable lengths are approximately equal.

Concerning the mass ratio, R has a relatively small effect on ω_1, but the high frequency ω_2 becomes more important to the double-pendulum motions for low payload-to-hook mass ratios. In brief, low payload-to-hook mass ratios and equal cable lengths are more representative to depict the double-pendulum motions of the crane for the point-to-point transport control with a constant $l_1 + l_2$.

According to eqs. (2.55), (2.56), and (2.57), Figure 2.6 shows the ratio of the high-mode contribution to the low-mode contribution for a large range of length and mass ratios, again assuming an overall length of 6 m. The surface indicates that double-pendulum effects cannot be ignored for systems with low payload-to-hook mass ratios. The second-mode contribution is particularly large when the suspension and rigging lengths are approximately equal.

2.4 Source Codes

Figures 2.5 and 2.6 can be numerically obtained by Matlab software [11]. The source codes are listed as follows.

Matlab codes to plot Figure 2.5

```
g=9.80;
R=linspace(0,2,25);
```

```
L2=linspace(6,0,25);
L1=6-L2;
[RR,LL2]=meshgrid(R,L2);
LL1=6.-LL2;
B=sqrt((RR+1).^2.*(1./LL1+1./LL2).^2-
  4.*((RR+1)./(LL1.*LL2)));
W2=sqrt(g/2)*sqrt((1+RR).*(1./LL1+1./LL2)+B);
W1=sqrt(g/2)*sqrt((1+RR).*(1./LL1+1./LL2)-B);

surf(LL2,RR,W2);
hold on;
surf(LL2,RR,W1);
xlabel('{\itl}_2 (m)');
ylabel('Mass ratio {\itR}');
zlabel('High and low frequencies');
```

Matlab codes to plot Figure 2.6

```
g=9.80;
R=linspace(0,1,25);
L2=linspace(1,4.8,25);
L1=6-L2;
X=L1./L2;
[XX,RR]=meshgrid(X,R);
LL2=6./(XX+1);
LL1=6.-LL2;
B=sqrt((RR+1).^2.*(1./LL1+1./LL2).^2-
  4.*((RR+1)./(LL1.*LL2)));
W1=sqrt(g/2)*sqrt((1+RR).*(1./LL1+1./LL2)-B);
W2=sqrt(g/2)*sqrt((1+RR).*(1./LL1+1./LL2)+B);
k=B.*LL1*g;
a=(-g*(1+RR))./(W1.^2.*W2.^2.*LL1.*LL2);
C1=(W1.*LL1.*(1+W2.^2.*a.*(LL1+LL2)))./k;
C2=-(W2.*LL1.*(1+W1.^2.*a.*(LL1+LL2)))./k;
Z=-C2./C1;
```

```
surf(XX,RR,Z);
xlabel('Length ratio, l_1/l_2');
ylabel('Mass ratio, m_1/m_2');
zlabel('Importance of low frequency, C_2/C_1');
```

References

[1] De León M, Rodrigues PR. Generalized classical mechanics and field theory: A geometrical approach of Lagrangian and Hamiltonian formalisms involving higher order derivatives. Amsterdam: Elsevier, 2011.

[2] Wang W, Yi JQ, Zhao DB, Liu DT. Design of a new stable sliding-mode controller of a class of second-order underactuated systems. IEE Proc Control Theory Appl, 2004, 151(6): 683–690.

[3] Qian DW, Tong SW, Yi JQ. Adaptive control based on incremental hierarchical sliding mode for overhead crane systems. Appl Math Inform Sci, 2013, 7(4): 1359–1364.

[4] Qian DW, Yi JQ, Zhao DB. How to automatically set an initial angle for balance control of a cart-pole system: An education case. Int J Elec Eng Educ, 2013, 50(1):57–68.

[5] Qian D, Yi J. Hierarchical sliding mode control for under-actuated cranes. Berlin: Springer, 2016.

[6] Liu DT, Guo WP, Yi JQ. Dynamics and GA-based stable control for a class of underactuated mechanical systems. Int J Control Autom Syst, 2008, 6(1): 35–43.

[7] O'Connor W, Habibi H. Gantry crane control of a double-pendulum, distributed-mass load, using mechanical wave concepts. Mech Sci, 2013, 4(2): 251–261.

[8] Singhose W, Kim D, Kenison M. Input shaping control of double-pendulum bridge crane oscillations. J Dyn Syst Meas Control-Trans. ASME, 2008, 130(3): DOI:10.1115/1.2907363.

[9] Mar R, Goyal A, Nguyen V, Yang TL, Singhose W. Combined input shaping and feedback control for double-pendulum systems. Mech Syst Signal Proc, 2017, 85: 267–277.

[10] Qian DW, Tong SW, Lee S. Fuzzy-logic-based control of payloads subjected to double-pendulum motion in overhead cranes. Autom Constr, 2016, 65: 133–143.

[11] Xue D, Chen Y, Atherton DP. Linear feedback control: Analysis and design with MATLAB. Philadelphia: Society for Industrial and Applied Mathematics, 2007.

3 Passivity-Based Anti-sway Control

Abstract: This chapter focuses on the anti-sway control problem of double-pendulum-type overhead cranes. The control method is passivity based. First, some basic concepts about passivity are presented. Then, taking double-pendulum-type overhead cranes into account, the passivity property of such systems is investigated. The system stability of the passivity-based method is proven. Some simulation results are illustrated to support the passivity-based anti-sway control. Finally, all the source codes are drawn.

Keywords: passivity, passivity-based control, stability, double-pendulum-type motions, overhead cranes

3.1 What is Passivity?

Passivity is one of the most physically appealing concepts in systems theory and has been used as a fundamental tool in the development of control tools for linear and nonlinear systems [1]. For robots and other mechanical systems, passivity is related to energy dissipation and leads to natural results on stability and tracking [2]. From the perspective of control theory, passivity provides us with a powerful tool for analysis of nonlinear system, which relates to Lyapunov stability [3].

Passivity is always concerned about energy. When a dynamical system is entitled passivity, its energy exchanged with the environment plays a pivotal role. In the passive system, the rate at which the energy flows into the system is not less than the increase in storage [4]. In other words, the passive system cannot store more energy than is supplied to it from outside, with the difference being the dissipated energy. The stored internal energy in the passive system is bounded by the externally supplied energy. It is well known that this energy dissipation property has important implications for closed-loop stability.

Additionally, the passivity property is preserved with respect to feedback and parallel interconnections of passive systems [5, 6]. This property is crucial in designing and analyzing highly networked systems. It is clear from the energy interpretation of passivity given above that it is intimately related with the physics of the system, and in particular its stability properties. For instance, viewing a feedback interconnection as a process of energy exchange, it is not surprising to know that passivity is invariant under negative feedback interconnection. In other words, a system composed of two passive subsystems is still passive if the two subsystems have feedback interconnection.

Despite the long history, passivity-based control is being actively utilized in addressing emerging problems. We will give a few basic definitions to start. Consider a

DOI 10.1515/9783110520101-003

dynamical system represented by the following state model:

$$\begin{aligned} \dot{\mathbf{x}} &= \mathbf{f}(\mathbf{x}, \mathbf{u}) \\ \mathbf{y} &= \mathbf{h}(\mathbf{x}, \mathbf{u}) \end{aligned} \tag{3.1}$$

In eq. (3.1), the system has the same number of inputs and outputs, p, $\mathbf{x} \in R^n$, $\mathbf{u} \in R^m$, $\mathbf{f} \in R^m$ is locally Lipschitz, $\mathbf{h} \in R^p$ is continuous, $\mathbf{f}(\mathbf{0}_n, \mathbf{0}_m) = \mathbf{0}_n$, $\mathbf{h}(\mathbf{0}_n, \mathbf{0}_m) = \mathbf{0}_p$, where $\mathbf{0}_n$, $\mathbf{0}_m$, and $\mathbf{0}_p$ are zero vectors, $\mathbf{0}_n \in R^n$, $\mathbf{0}_m \in R^m$, and $\mathbf{0}_p \in R^p$.

System (3.1) is said to be

$$\text{passive if } \dot{S} \le \mathbf{u}^T\mathbf{y} \quad \text{for all } (\mathbf{x}, \mathbf{u}) \in R^n \times R^p \tag{3.2}$$

$$\text{strictly passive if } \dot{S} \le \mathbf{u}^T\mathbf{y} - \varphi(\mathbf{x}) \quad \text{for } \varphi > 0 \tag{3.3}$$

$$\text{input passive if } \dot{S} \le \mathbf{u}^T\mathbf{y} - \mathbf{u}^T\psi(\mathbf{u}) \quad \text{for } \mathbf{u}^T\psi(\mathbf{u}) > 0 \tag{3.4}$$

$$\text{output passive if } \dot{S} \le \mathbf{u}^T\mathbf{y} - \mathbf{y}^T\rho(\mathbf{y}) \quad \text{for } \mathbf{y}^T\rho(\mathbf{y}) > 0 \tag{3.5}$$

Here S: $R^n \to R^p$, called the storage function, is a semidefinite scalar function.

According to the definition of passivity in eq. (3.1), the passivity-based control strongly relies on the system model. However, for various practical applications obtaining a precise model of the system is extremely difficult [7]. This drawback may restrict the applications of the passivity-based control synthesis. Fortunately, the double-pendulum-type overhead crane systems can be theoretically modeled by Lagrangian mechanics. As a result, it is possible to investigate the passivity-based control for the double-pendulum-type overhead crane systems. Without loss of generality, the two-dimensional case is taken into consideration.

3.2 Passivity Analysis

Recall the double-pendulum-type overhead crane model in the form of two dimensions (2.31):

$$\mathbf{M}(\mathbf{q})\,\ddot{\mathbf{q}} + \mathbf{C}(\mathbf{q}, \dot{\mathbf{q}})\,\dot{\mathbf{q}} + \mathbf{G}(\mathbf{q}) = \boldsymbol{\tau} \tag{3.6}$$

All the notations and parameters in eq. (3.6) have been defined in eq. (2.31), and they are not explained again. $\mathbf{M}(\mathbf{q})$ has been formulated in Chapter 2. Apparently, it is a positive-definite symmetric matrix.

Recall the Lagrangian of the crane system in eq. (2.17). Its expression can have another form of

$$L = K - P = \frac{1}{2}\dot{\mathbf{q}}^T\mathbf{M}(\mathbf{q})\dot{\mathbf{q}} - P \tag{3.7}$$

In eq. (3.7), $\mathbf{M}(\mathbf{q})$ is the inertia matrix, also called inertia tensor. According to the definition of the products of inertia, the inertia matrix $\mathbf{M}(\mathbf{q})$ is always positive definite and symmetric.

Lemma 3.1 ([1]). *Concerning eq. (3.6), define*

$$\mathbf{N}(\mathbf{q},\dot{\mathbf{q}}) = \dot{\mathbf{M}}(\mathbf{q}) - 2\mathbf{C}(\mathbf{q},\dot{\mathbf{q}}) \tag{3.8}$$

then $\mathbf{N}(\mathbf{q},\dot{\mathbf{q}})$*is a skew-symmetric matrix.*

Proof. Each element of the derivative of the inertia matrix $\mathbf{M}(\mathbf{q})$ can be given in the following form:

$$\dot{m}_{ij}(\mathbf{q}) = \sum_{k=1}^{3} \frac{\partial m_{ij}(\mathbf{q})}{\partial q_k}\dot{q}_k \tag{3.9}$$

Then, each element of $\mathbf{N}(\mathbf{q},\dot{\mathbf{q}})$ in eq. (3.8) can have the form

$$n_{ij}(\mathbf{q},\dot{\mathbf{q}}) = \dot{m}_{ij}(\mathbf{q}) - 2c_{ij}(\mathbf{q}) = \sum_{k=1}^{3}\left(\frac{\partial m_{kj}(\mathbf{q})}{\partial q_i} - \frac{\partial m_{ik}(\mathbf{q})}{\partial q_j}\right)\dot{q}_k \tag{3.10}$$

Since the inertia matrix $\mathbf{M}(\mathbf{q})$ is always positive definite and symmetric, it is straightforward to deduce that

$$n_{ji}(\mathbf{q},\dot{\mathbf{q}}) = \sum_{k=1}^{3}\left(\frac{\partial m_{ki}(\mathbf{q})}{\partial q_j} - \frac{\partial m_{jk}(\mathbf{q})}{\partial q_i}\right)\dot{q}_k = -\sum_{k=1}^{3}\left(\frac{\partial m_{kj}(\mathbf{q})}{\partial q_i} - \frac{\partial m_{ik}(\mathbf{q})}{\partial q_j}\right)\dot{q}_k = -n_{ij}(\mathbf{q},\dot{\mathbf{q}}) \tag{3.11}$$

that is eq. (3.11) indicates $\mathbf{N}(\mathbf{q},\dot{\mathbf{q}})$ is a skew-symmetric matrix.

Lemma 3.2 ([8]). *The double-pendulum-type overhead crane system formulated by eq. (3.8) is a passive system.*

Proof. The total energy of the overhead crane system can be determined by its kinetic energy plus its potential energy:

$$E(\mathbf{q},\dot{\mathbf{q}}) = \frac{1}{2}\dot{\mathbf{q}}^T\mathbf{M}(\mathbf{q})\dot{\mathbf{q}} + P(\mathbf{q}) \tag{3.12}$$

where $\frac{1}{2}\dot{\mathbf{q}}^T\mathbf{M}(\mathbf{q})\dot{\mathbf{q}}$ and $P(\mathbf{q})$ denote the kinetic energy and the potential energy, respectively.

Select a reference point of the potential energy such that $P(\mathbf{q}) \geq 0$. Then, it can be concluded that $E(\mathbf{q}, \dot{\mathbf{q}}) \geq 0$. Differentiating $E(\mathbf{q}, \dot{\mathbf{q}})$ in eq. (3.12) with respect to time t yields

$$\dot{E}(\mathbf{q}, \dot{\mathbf{q}}) = \dot{\mathbf{q}}^T \mathbf{M}(\mathbf{q})\ddot{\mathbf{q}} + \frac{1}{2}\dot{\mathbf{q}}^T \dot{\mathbf{M}}(\mathbf{q})\dot{\mathbf{q}} + \dot{\mathbf{q}}^T \frac{\partial P(\mathbf{q})}{\partial \mathbf{q}} \tag{3.13}$$

Due to $\frac{\partial P(\mathbf{q})}{\partial \mathbf{q}} = \mathbf{G}(\mathbf{q})$, eq. (3.13) can be rearranged by

$$\dot{E}(\mathbf{q}, \dot{\mathbf{q}}) = \dot{\mathbf{q}}^T \left[\mathbf{M}(\mathbf{q})\ddot{\mathbf{q}} + \mathbf{G}(\mathbf{q})\right] + \frac{1}{2}\dot{\mathbf{q}}^T \dot{\mathbf{M}}(\mathbf{q})\dot{\mathbf{q}} \tag{3.14}$$

Substituting eq. (3.6) into eq. (3.14) gives

$$\dot{E}(\mathbf{q}, \dot{\mathbf{q}}) = \dot{\mathbf{q}}^T \boldsymbol{\tau} + \frac{1}{2}\dot{\mathbf{q}}^T \left[\dot{\mathbf{M}}(\mathbf{q}) - 2\mathbf{C}(\mathbf{q}, \dot{\mathbf{q}})\right] \dot{\mathbf{q}} = \dot{\mathbf{q}}^T \boldsymbol{\tau} + \frac{1}{2}\dot{\mathbf{q}}^T \mathbf{N}(\mathbf{q})\dot{\mathbf{q}} \tag{3.15}$$

According to Lemma 3.1, $\mathbf{N}(\mathbf{q})$ is a skew-symmetric matrix; hence, $\dot{\mathbf{q}}^T \mathbf{N}(\mathbf{q})\dot{\mathbf{q}} = 0$ and eq. (3.15) can be written as

$$\dot{E}(\mathbf{q}, \dot{\mathbf{q}}) = \dot{\mathbf{q}}^T \boldsymbol{\tau} \tag{3.16}$$

The mechanical structure of the double-pendulum-type overhead crane system has three degrees of freedom but has only one input. The system is a typical underactuated system with the single input. In eq. (2.31), $\mathbf{q} = [x, \theta_1, \theta_2]^T$ and $\boldsymbol{\tau} = [f, 0, 0]^T$. Finally, eq. (3.16) can have the form

$$\dot{E}(\mathbf{q}, \dot{\mathbf{q}}) = \dot{x}f \tag{3.17}$$

Integrating both sides of eq. (3.17) yields

$$\int_0^t \dot{x}f\mathrm{d}t = \int_0^t \dot{E}(\mathbf{q}, \dot{\mathbf{q}})\mathrm{d}t = E(\mathbf{q}(t), \dot{\mathbf{q}}(t)) - \dot{E}(\mathbf{q}(0), \dot{\mathbf{q}}(0)) \tag{3.18}$$

Recall that $E(\mathbf{q}(t), \dot{\mathbf{q}}(t)) \geq 0$ holds true. Hence, eq. (3.18) can be rearranged as

$$\int_0^t \dot{x}f\mathrm{d}t \geq -\dot{E}(\mathbf{q}(0), \dot{\mathbf{q}}(0)) \tag{3.19}$$

that is, $\langle f|\dot{x}\rangle_t \geq -\dot{E}(\mathbf{q}(0), \dot{\mathbf{q}}(0))$.

Concerning the definition about passivity in eq. (3.2), eq. (3.19) indicates that the double-pendulum-type overhead crane system is passive with respect to input f and output $\dot{x}$. As a result, it is possible to develop an anti-sway controller via the property of passivity.

3.3 Control Design Based on Passivity

To develop a passivity-based controller in the sense of Lyapunov, the following Lyapunov candidate is picked up for the double-pendulum-type overhead crane system:

$$V(\mathbf{q}, \dot{\mathbf{q}}) = k_E(E(\mathbf{q}, \dot{\mathbf{q}}) - P(\mathbf{q}_d)) + \frac{1}{2}k_D\dot{x}^2 + \frac{1}{2}k_P(x - x_d)^2 \tag{3.20}$$

where x_d is the control objective of the actuated part, indicating the desired position of the trolley; k_E, k_D, and k_P are positive constants; $P(\mathbf{q}_d)$ is the potential energy at the desired position.

Concerning the anti-sway control task, the crane system should have no oscillations at the desired position, which means that the system at the desired position has the minimal potential energy point among all the accessible states. Thus, the selected Lyapunov candidate is not negative, that is, $V(\mathbf{q}, \dot{\mathbf{q}}) \geq 0$.

Differentiating $V(\mathbf{q}, \dot{\mathbf{q}})$ in eq. (3.20) with respect to time t yields

$$\dot{V}(\mathbf{q}, \dot{\mathbf{q}}) = k_E\dot{E}(\mathbf{q}, \dot{\mathbf{q}}) + k_D\dot{x}\ddot{x} + k_P\dot{x}(x - x_d) \tag{3.21}$$

Substituting eq. (3.17) into eq. (3.21) gives

$$\dot{V}(\mathbf{q}, \dot{\mathbf{q}}) = k_E\dot{x}f + k_D\dot{x}\ddot{x} + k_P\dot{x}(x - x_d) = \dot{x}[k_Ef + k_D\ddot{x} + k_P(x - x_d)] \tag{3.22}$$

In order to ensure the closed-loop stability in the sense of Lyapunov, it is expected that

$$\dot{V}(\mathbf{q}, \dot{\mathbf{q}}) = -k\dot{x}^2 \tag{3.23}$$

where k is a positive constant. From eqs. (3.22) and (3.23), the following equation can be drawn:

$$k_Ef + k_D\ddot{x} + k_P(x - x_d) = -k\dot{x} \tag{3.24}$$

Further, $\ddot{x}$ in eq. (3.24) can be written as

$$\ddot{x} = \mathrm{Z}\ddot{\mathbf{x}} \tag{3.25}$$

Here $\mathrm{Z} = \begin{bmatrix} 1 & 0 & 0 \end{bmatrix}$.

Substituting eq. (3.6) into eq. (3.25) yields

$$\ddot{x} = \mathrm{Z}\mathbf{M}^{-1}(\mathbf{q})\left(\boldsymbol{\tau} - \mathbf{C}(\mathbf{q}, \dot{\mathbf{q}})\,\dot{\mathbf{q}} - \mathbf{G}(\mathbf{q})\right) \tag{3.26}$$

Further, $f = \mathrm{Z}\boldsymbol{\tau}$. Therefore, $\boldsymbol{\tau}$ can be written as

$$\boldsymbol{\tau} = \mathrm{Z}^Tf \tag{3.27}$$

Replacing $\boldsymbol{\tau}$ in eq. (3.26) by eq. (3.27) gives

$$\ddot{x} = \mathrm{Z}\mathbf{M}^{-1}(\mathbf{q})\left(\mathrm{Z}^T f - \mathbf{C}(\mathbf{q}, \dot{\mathbf{q}})\,\dot{\mathbf{q}} - \mathbf{G}(\mathbf{q})\right) \tag{3.28}$$

Substituting eq. (3.28) into eq. (3.24) yields

$$f = \left(k_E + k_D \mathrm{Z}\mathbf{M}^{-1}(\mathbf{q})\,\mathrm{Z}^T\right)^{-1}\left\{k_D \mathrm{Z}\mathbf{M}^{-1}(\mathbf{q})\left[\mathbf{C}(\mathbf{q}, \dot{\mathbf{q}})\,\dot{\mathbf{q}} + \mathbf{G}(\mathbf{q})\right] - k_P(x - x_d) - k\dot{x}\right\} \tag{3.29}$$

Lemma 3.3 (Barbalat's lemma [3]). *Suppose* $f(t) \in R^1(a, \infty)$ *and* $\lim_{t\to\infty} f(t) = \alpha$ *where* $\alpha < \infty$. *If the derivative of* $f(t)$ *is uniformly continuous, then* $\lim_{t\to\infty} \dot{f}(t) = 0$.

Proof. This lemma can be proven by contradiction.

Suppose $\lim_{t\to\infty} \dot{f}(t) \neq 0$. Then $\exists \varepsilon > 0$ and a monotone-increasing sequence $\{t_n\}$ such that $t_n \to \infty$ as $n \to \infty$ and $\left|\dot{f}(t_n)\right| \geq \varepsilon$ for all $n \in N$.

Since $\dot{f}(t)$ is uniformly continuous, for such ε, $\exists \delta > 0$ such that $\forall n \in N$

$$\left|t - t_n\right| < \delta \Rightarrow \left|\dot{f}(t) - \dot{f}(t_n)\right| < \frac{\varepsilon}{2} \tag{3.30}$$

Hence, if $t \in [t_n, t_n + \delta]$, then

$$\begin{aligned} \left|\dot{f}(t)\right| &= \left|\dot{f}(t_n) - [\dot{f}(t_n) - \dot{f}(t)]\right| \\ &\geq \left|\dot{f}(t_n)\right| - \left|\dot{f}(t_n) - \dot{f}(t)\right| \\ &\geq \varepsilon - \frac{\varepsilon}{2} = \frac{\varepsilon}{2} \end{aligned} \tag{3.31}$$

Then since $f(t) \in R^1$, there exists

$$\begin{aligned} \left|\int_a^{t_n+\delta} \dot{f}(t)\mathrm{d}t - \int_a^{t_n} \dot{f}(t)\mathrm{d}t\right| &= \left|\int_{t_n}^{t_n+\delta} \dot{f}(t)\mathrm{d}t\right| \geq \int_{t_n}^{t_n+\delta} \left|\dot{f}(t)\right| \mathrm{d}t \\ &\geq \int_{t_n}^{t_n+\delta} \frac{\varepsilon}{2}\mathrm{d}t = \frac{\varepsilon\delta}{2} > 0 \end{aligned} \tag{3.32}$$

However,

$$\begin{aligned} \lim_{t\to\infty}\left|\int_a^{t_n+\delta} \dot{f}(t)\mathrm{d}t - \int_a^{t_n} \dot{f}(t)\mathrm{d}t\right| &= \lim_{t\to\infty}\left|f(t_n + \delta) - f(t_n)\right| \\ &= \left|\lim_{t\to\infty} f(t_n + \delta) - \lim_{t\to\infty} f(t_n)\right| \\ &= |\alpha| - |\alpha| = 0 \end{aligned} \tag{3.33}$$

This is a contradiction. Therefore, $\lim_{t\to\infty} \dot{f}(t) = 0$.

Theorem 3.1 ([8]). Consider the double-pendulum-type overhead crane system (3.6). Its equilibrium $\mathbf{q}_d$ has the minimal potential energy in the system-accessible space. If its controller is designed by eq. (3.29), then the controller can make the closed-loop system converge to the equilibrium point $\mathbf{q}_d$ or the stable trajectory $\frac{1}{2}\dot{\mathbf{q}}^T\mathbf{M}(\mathbf{q})\dot{\mathbf{q}} + \mathbf{P}(\mathbf{q}) = C$, where C is a positive constant.

Proof. From eqs. (3.20) and (3.23), $V(\mathbf{q}, \dot{\mathbf{q}}) \geq 0$ and $\dot{V}(\mathbf{q}, \dot{\mathbf{q}}) \leq 0$ hold true. This fact can be divided into two cases.

Case 1. When $\mathbf{q} = \mathbf{q}_d$ and $\dot{\mathbf{q}} = 0$ hold true at the same time. Then $V(\mathbf{q}, \dot{\mathbf{q}}) = 0$. From eq. (3.29), the control law $f = 0$, indicating the crane system is stabilized at the desired position.

Case 2. When $\mathbf{q} \neq \mathbf{q}_d$ and $\dot{\mathbf{q}} \neq 0$, $V(\mathbf{q}, \dot{\mathbf{q}})$ is a positive-definite function. From eq. (3.23), $\dot{V}(\mathbf{q}, \dot{\mathbf{q}}) \leq 0$ holds true. Therefore,

$$\dot{x} \in L_2 \cap L_\infty \tag{3.34}$$

Further, $E(\mathbf{q}, \dot{\mathbf{q}})$ and x are bounded. From definition (3.12), $\mathbf{q}$, $\dot{\mathbf{q}}$, and $\mathbf{M}(\mathbf{q})$ are bounded. Since $\mathbf{M}(\mathbf{q})$ is a positive-definite symmetric matrix, $\mathbf{M}^{-1}(\mathbf{q})$ is also bounded. From the crane dynamics (3.6), $\ddot{\mathbf{q}}$ is bounded as well, the formulas mean that $\ddot{V}(\mathbf{q}, \dot{\mathbf{q}}) = -2k\dot{x}\ddot{x}$ is bounded. Thus, the function $\dot{V}(\mathbf{q}, \dot{\mathbf{q}})$ is uniformly continuous.

According to Barbalat's lemma, we can obtain

$$\lim_{t\to\infty} \dot{V}(\mathbf{q}, \dot{\mathbf{q}}) = 0, \text{ i.e., } \lim_{t\to\infty} \dot{x} = 0 \tag{3.35}$$

It can be concluded that $\lim_{t\to\infty} V(\mathbf{q}, \dot{\mathbf{q}})$ and $\lim_{t\to\infty}(x - x_d)$ are constant. From eqs. (3.17) and (3.13), we have

$$\lim_{t\to\infty} E(\mathbf{q}, \dot{\mathbf{q}}) \text{ is constant and } \lim_{t\to\infty} f \text{ is constant} \tag{3.36}$$

If $\lim_{t\to\infty} f \neq 0$, then x will change with respect to $t \to \infty$. The fact is inconsistent with $\lim_{t\to\infty}(x - x_d)$ and is constant. Thus $\lim_{t\to\infty} f = 0$ is true, indicating that

$$\lim_{t\to\infty} x = x_d \tag{3.37}$$

The following stability analysis can be divided into two parts.

Part 1: When $\lim_{t\to\infty} E(\mathbf{q}, \dot{\mathbf{q}}) = P(\mathbf{q}_d)$, eq. (3.12) can be written as

$$\frac{1}{2}\dot{\mathbf{q}}^T\mathbf{M}(\mathbf{q})\dot{\mathbf{q}} + (P(\mathbf{q}) - P(\mathbf{q}_d)) = 0 \text{ as } t \to \infty \tag{3.38}$$

In eq. (3.28), both the two terms on the left-hand side are greater than or equal to zero. Therefore, kinetic energy of the system is equal to zero, and potential energy of the system is equal to the potential energy at the desired position, that is, $\mathbf{q} = \mathbf{q}_d$ and

$\dot{\mathbf{q}} = 0$. The fact indicates that the overhead crane system is stabilized at the desired position.

Part 2: When $\lim_{t\to\infty} E(\mathbf{q}, \dot{\mathbf{q}}) = C \neq P(\mathbf{q}_d)$, the system converges to the following stable trajectory:

$$\frac{1}{2}\dot{\mathbf{q}}^T\mathbf{M}(\mathbf{q})\dot{\mathbf{q}} + P(\mathbf{q}) = C \text{ as } t \to \infty \tag{3.39}$$

Here C is a greater than $P(\mathbf{q}_d)$ and it can be calculated by eqs. (3.39) and (3.6).

3.4 Simulation Results

To verify the feasibility and validity of the passivity-based anti-sway control method for the transport control problem, some numerical simulations are carried out in this section. The control objective is to transport the payload to a desired position. Without loss of generality, the desired position x_d is located at 1 m, that is, $x_d = 1$ m from the origin. Meanwhile, the controller needs to suppress the double-pendulum motions of the hook and payload. The simulation results are presented in Figures 3.1–3.8.

The physical parameters of the double-pendulum-type crane system are determined by $m = 0.5$ kg, $m_1 = 0.25$ kg, $m_2 = 0.5$ kg, $l_1 = 0.5$ m, and $l_2 = 0.5$ m, which are regarded as nominal parameters. Note that the two lengths are equal for this crane. Recall Chapter 2. From Figure 2.6, the contribution of the low-frequency oscillation to problematic swing amplitude is particularly large for a constant $l_1 + l_2$ when the two

Figure 3.1: Trolley position versus time by the passivity-based anti-sway control.

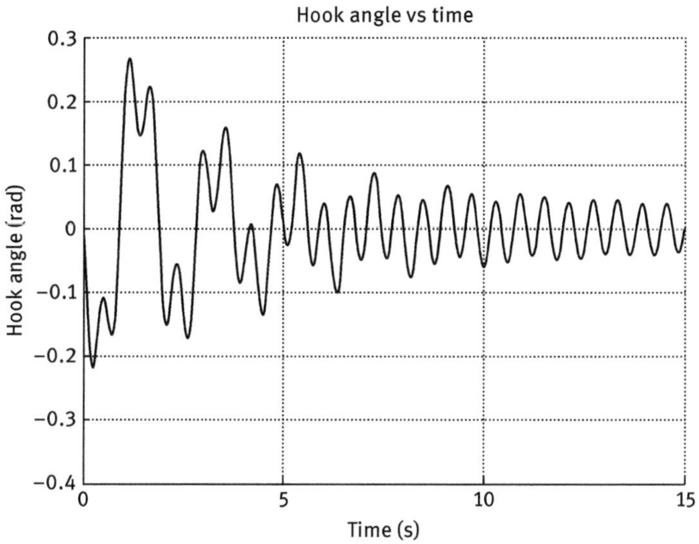

Figure 3.2: Hook angle versus time by the passivity-based anti-sway control.

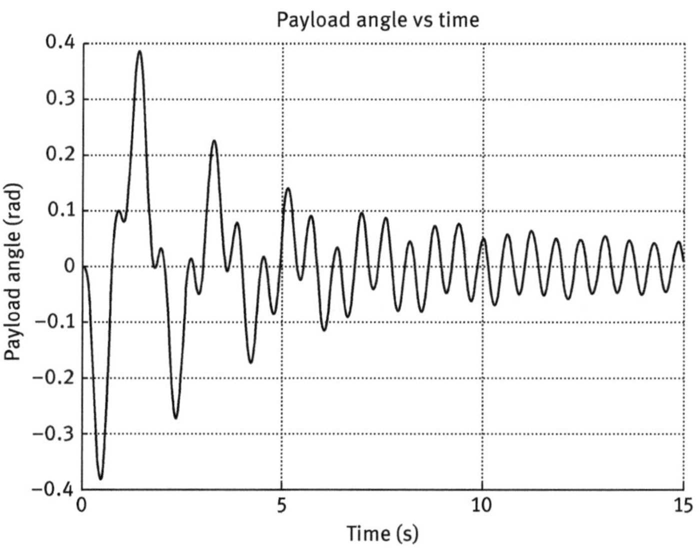

Figure 3.3: Payload angle versus time by the passivity-based anti-sway control.

cable lengths are approximately equal. In order to testify the passivity-based anti-sway controller, both of the cable lengths are equal.

From eq. (3.29), some controller parameters should be set. However, it is very difficult to determine the parameters because these parameters have no obvious physics meaning and there are complex couplings among them. For the purpose of

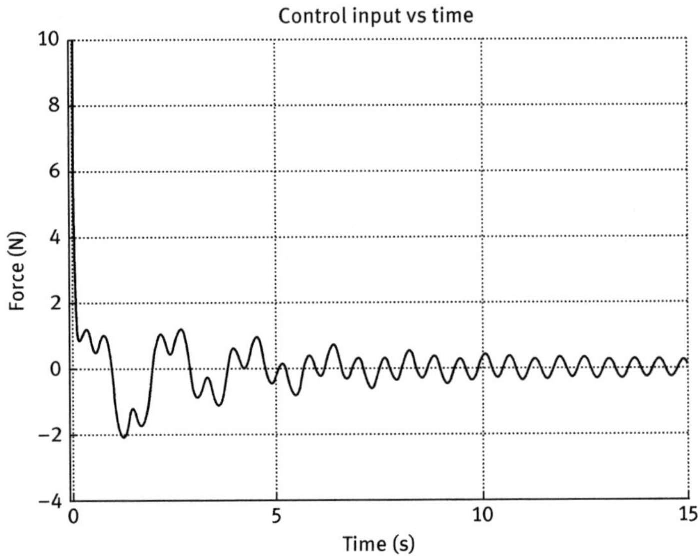

Figure 3.4: Control input versus time by the passivity-based anti-sway control.

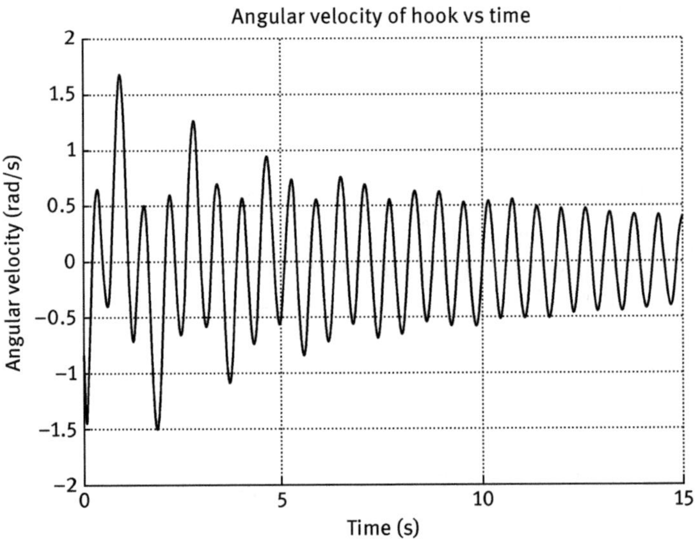

Figure 3.5: Angular velocity of hook versus time by the passivity-based anti-sway control.

simplification, only the position term in eq. (9) is taken into consideration, that is, $k_d = 0$ in (3.29). Finally, the passivity-based controller becomes

$$f = -k_1\dot{x} - k_2(x - x_d) \tag{3.40}$$

where $k_1 = \frac{k}{k_E}$ and $k_2 = \frac{k_P}{k_E}$.

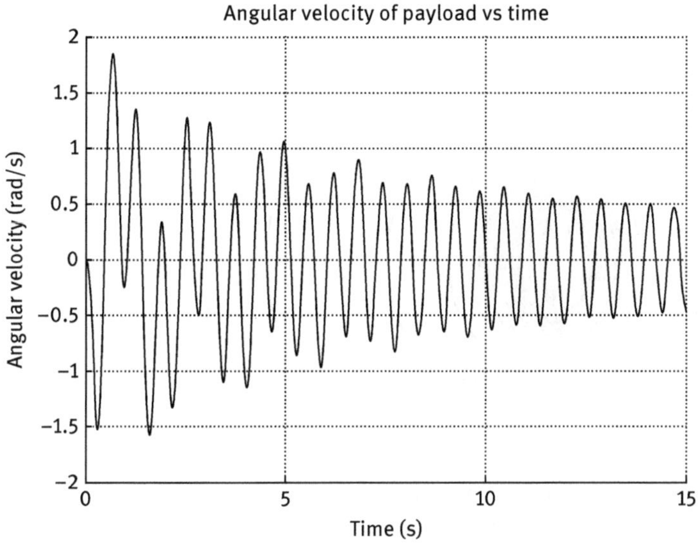

Figure 3.6: Angular velocity of payload versus time by the passivity-based anti-sway control.

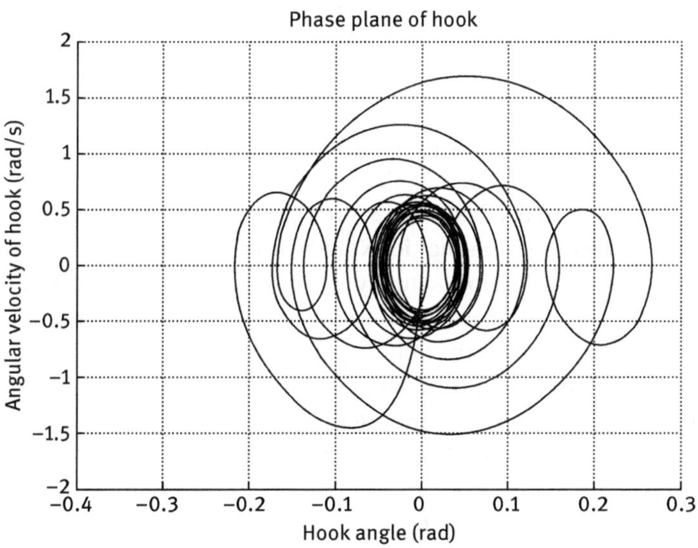

Figure 3.7: Phase plane of hook.

We know that the performance of a control system is concerned about the controller parameters. In order to optimize the designed passivity-based anti-sway control system, its controller parameters are tuned by trial and error until its best control performance is achieved. Finally, k_1 and k_2 in eq. (3.40) are determined as

$$k_1 = 10 \ldots k_2 = 10 \tag{3.41}$$

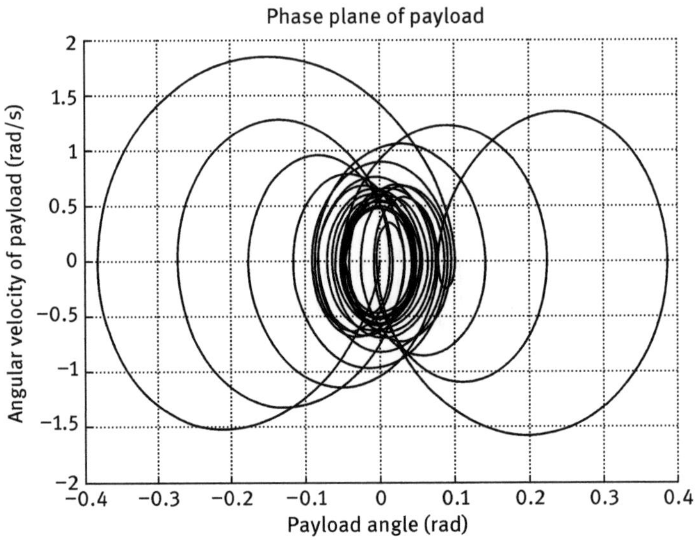

Figure 3.8: Phase plane of payload.

As shown in Figure 3.1, the passivity-based anti-sway controller is able to achieve the transport control objective. The trolley is stabilized at the desired position, that is, $x_d = 1$ m.

The angles of payload and hook are illustrated in Figures 3.2 and 3.3. Their angular velocities are shown in Figures 3.5 and 3.6. The curves indicate that the payload and hook are not completely stabilized at the desired position. They have slight oscillations and the oscillations seem harmonic. Note that the crane dynamics are ideal without any friction. In fact, friction actually exists, which can dissipate oscillations. Consequently, the passivity-based anti-sway controller can suppress the double-pendulum-type motions, stabilize the double-pendulum-type payload, as well as accomplish the transport control task. However, the control method needs a long time to dissipate oscillations. Although the controller is very simple and the control system has guaranteed stability, its control performance still has some room to be improved.

Further, the control force applied to the trolley is displayed in Figure 3.4. From Figure 3.4, the overhead crane system needs a large control force at the outset. This may overload the driven motor in reality.

In order to describe the double-pendulum-type motions, the phase planes of the hook and payload are demonstrated in Figures 3.7 and 3.8. Recall Figures 3.5 and 3.6. The curves verify that the payload and hook have slight oscillations when the trolley arrives at the desired position. The phenomenon inspires our motivation to develop other anti-sway control methods for better performance.

3.5 Source Codes

Figures 3.1–3.8 can be numerically obtained by Matlab software [9]. The source codes are listed as follows:

```
function passivecontrol
clear all;
close all;

%crane parameters
len1=0.5;
len2=0.5;
mass=0.5;
mass1=0.25;
mass2=0.5;
g=9.81;

%initial values
s=0;
v=0;
th1=0;
th2=0;
w1=0;
w2=0;
f=0;
t=0;
s_1=s;
v_1=v;
th_1=th1;
th_2=th2;
w_1=w1;
w_2=w2;
dt=0.01; %sample 0.01s
xend=1; %constraint of trolley position [-1 1]
```

```
%storage of the states
sdata=s;
vdata=v;
th1data=th1;
th2data=th2;
w1data=w1;
w2data=w2;
fdata=f;
tdata=t;

for i=1:1:1500
s=s_1;
v=v_1;
th1=th_1;
th2=th_2;
w1=w_1;
w2=w_2;

M=[mass+mass1+mass2 (mass1+mass2)*len1*cos(th1)
        mass2*len2*cos(th2);...
        (mass1+mass2)*len1*cos(th1) (mass1+mass2)*len1^2
        mass2*len1*len2*cos(th1-th2);...
        mass2*len2*cos(th2) mass2*len1*len2*cos(th1-th2)
        mass2*len2^2];

C=[0 -1*(mass1+mass2)*len1*w1*sin(th1) -1*
        mass2*len2*w2*sin(th2);...
        0 0 mass2*len1*len2*w1*sin(th1-th2);...
        0 -1*mass2*len1*len2*w1*sin(th1-th2) 0];
G=[0;(mass1+mass2)*g*len1*sin(th1);mass2*g*len2*sin(th2)];

f=controller(xend,s,v);
T=[f;0;0];
```

```
x=inv(M)*(T-G-C*[v;w1;w2]);
y=[v;w1;w2]+dt*x;
v_1=y(1);
w_1=y(2);
w_2=y(3);
z=[s;th1;th2]+0.5*dt^2*x+dt*[v;w1;w2];
s_1=z(1);
th_1=z(2);
th_2=z(3);
t=t+dt;
tdata = [tdata,t];
fdata = [fdata,f];
vdata = [vdata,v_1];
th1data=[th1data,th_1];
th2data=[th2data,th_2];
w1data=[w1data,w_1];
w2data=[w2data,w_2];
sdata=[sdata,s_1];
end

%plot curves
h1=figure(1);
set(h1,'Position',[100,200,500,300],...
          'Menu','none',...
          'NumberTitle','off',...
          'Name','Trolley Position'...
);
axes1=axes('Parent',h1);
title(axes1,'Trolley position vs time','FontSize',15);
xlabel(axes1,'Time (s)');
ylabel(axes1,'Trolley position£¨ m£©');
hold(axes1,'all');
plot(tdata,sdata);
grid on;
```

```
h2=figure(2);
set(h2,'Position',[800,200,500,300],...
          'Menu','none',...
          'NumberTitle','off',...
          'Name','Hook angle'...
);
axes2=axes('Parent',h2);
title(axes2,'Hook angle vs time','FontSize',15);
xlabel(axes2,'Time(s)');
ylabel(axes2,'Hook angle(rad)');
hold(axes2,'all');
plot(tdata,th1data,'r');
grid on;

h3=figure(3);
set(h3,'Position',[800,200,500,300],...
          'Menu','none',...
          'NumberTitle','off',...
          'Name','Payload angle'...
);
axes3=axes('Parent',h3);
title(axes3,'Payload angle vs time','FontSize',15);
xlabel(axes3,'Time(s)');
ylabel(axes3,'Payload angle(rad)');
hold(axes3,'all');
plot(tdata,th2data,'r');
grid on;

h4=figure(4);
set(h4,'Position',[800,200,500,300],...
          'Menu','none',...
          'NumberTitle','off',...
          'Name','Control input'...
);
```

```
axes4=axes('Parent',h4);
title(axes4,'Control input vs time','FontSize',15);
xlabel(axes4,'Time(s)');
ylabel(axes4,'Force(N)');
hold(axes4,'all');
plot(tdata,fdata,'r');
grid on;

h5=figure(5);
set(h5,'Position',[800,200,500,300],...
          'Menu','none',...
          'NumberTitle','off',...
          'Name','Angular velocity of hook'...
);
axes5=axes('Parent',h5);
title(axes5,'Angular velocity of hook vs time',
'FontSize',15);
xlabel(axes5,'Time(s)');
ylabel(axes5,'Angular velocity (rad/s)');
hold(axes5,'all');
plot(tdata,w1data,'r');
grid on;

h6=figure(6);
set(h6,'Position',[800,200,500,300],...
          'Menu','none',...
          'NumberTitle','off',...
          'Name','Angular velocity of payload'...
);
axes6=axes('Parent',h6);
title(axes6,'Angular velocity of payload vs time',
'FontSize',15);
xlabel(axes6,'Time(s)');
ylabel(axes6,'Angular velocity(rad/s)');
```

```
hold(axes6,'all');
plot(tdata,w2data,'r');
grid on;

h7=figure(7);
set(h7,'Position',[800,200,500,300],...
          'Menu','none',...
          'NumberTitle','off',...
          'Name','Angular velocity of payload'...
);
axes7=axes('Parent',h7);
title(axes7,'Phase plane of payload','FontSize',15);
xlabel(axes7,'Hook angle(rad)');
ylabel(axes7,'Angular velocity of hook(rad/s)');
hold(axes7,'all');
plot(th1data,w1data,'r');
grid on;

h8=figure(8);
set(h8,'Position',[800,200,500,300],...
          'Menu','none',...
          'NumberTitle','off',...
          'Name','Angular velocity of payload'...
);
axes8=axes('Parent',h8);
title(axes8,'Phase plane of payload','FontSize',15);
xlabel(axes8,'Hook angle(rad)');
ylabel(axes8,'Angular velocity of hook(rad/s)');
hold(axes8,'all');
plot(th2data,w2data,'r');
grid on;

function f = controller( xend,s,v )
%UNTITLED Summary of this function goes here
```

```
% Detailed explanation goes here
k1=10;
k2=10;
f=k1*(xend-s)-k2*v;
```

References

[1] Ortega R, Perez JA, Nicklasson PJ, Sira-Ramirez H. Passivity-based control of Euler-Lagrange systems: Mechanical, electrical and electromechanical applications. Berlin: Springer, 2013.

[2] Fantoni I, Lozano R, Spong MW. Energy based control of the pendubot. IEEE Trans Autom Control, 2000, 45(4): 725–729.

[3] Khalil HK. Nonlinear systems. Upper Saddle River, NJ: Prentice-Hall, 1996.

[4] Spong MW, Holm JK, Lee D. Passivity-based control of bipedal locomotion. IEEE Robot Autom Mag, 2007, 14(2): 30–40.

[5] Ortega R, Van Der Schaft A, Maschke B, Escobar G. Interconnection and damping assignment passivity-based control of port-controlled Hamiltonian systems. Automatica, 2002, 38(4): 585–596.

[6] Gao H, Chen T, Chai T. Passivity and passification for networked control systems. SIAM J. Control Optim, 2007, 46(4): 1299–1322.

[7] Qian DW, Tong SW, Lee S. Fuzzy-logic-based control of payloads subjected to double-pendulum motion in overhead cranes. Autom Constr, 2016, 65: 133–143.

[8] Liu DT, Guo WP, Yi JQ. Dynamics and GA-based stable control for a class of underactuated mechanical systems. Int J Control Autom Syst, 2008, 6(1): 35–43.

[9] Xue D, Chen Y, Atherton DP. Linear feedback control: analysis and design with MATLAB. Philadelphia, PA: Society for Industrial and Applied Mathematics, 2007.

4 Sliding-Mode-Based Anti-sway Control

Abstract: This chapter focuses on the methodology of sliding mode for the anti-sway control problem of double-pendulum-type overhead cranes. The methodology of sliding mode is a class of design tools, which contains some basic design methods and their derivatives. This chapter starts with some basic concepts about sliding mode. Then, two advanced sliding mode methods, such as incremental sliding mode control and hierarchical sliding mode control, are presented for double-pendulum-type overhead cranes. The stability analyses are also given in the sense of Lyapunov. Simulation results are illustrated to support the sliding-mode-based anti-sway control methods. Finally, all the source codes are drawn.

Keywords: sliding mode control, incremental sliding mode, hierarchical sliding mode, double-pendulum-type motions, overhead cranes

4.1 Sliding Mode Control

Sliding mode control is also named sliding mode variable structure control or variable structure control with sliding mode [1, 2]. From the two aliases, sliding mode control is often associated with variable structure control and it has a close relationship with variable structure control. To understand about the sliding mode control, some background information about variable structure control is presented.

Variable structure control has a long history. It appeared worldwide till it was reported by V.I. Utkin [3]. Suggested by its name, a variable structure control system has a control law that deliberately changed during the control process based on some predefined rules.

For the purpose of illustration, consider a second-order system:

$$\ddot{x}(t) = u(t) \tag{4.1}$$

Design the following feedback control law:

$$u(t) = -kx(t) \tag{4.2}$$

Here $k > 0$.

A closed-loop system comprises eqs. (4.1) and (4.2). The phase plane of such a closed-loop system can be described by $\dot{x}$ versus x. The plot can be demonstrated as an ellipse. As shown in Figure 4.1, the shape completely depends on the initial conditions. Figure 4.1 illustrates the system trajectory under the initial conditions $x(0) = 1$ and $\dot{x}(0) = 0$, where $k = 0.5$ in Figure 4.1(a) and $k = 5.5$ in Figure 4.1(b).

DOI 10.1515/9783110520101-004

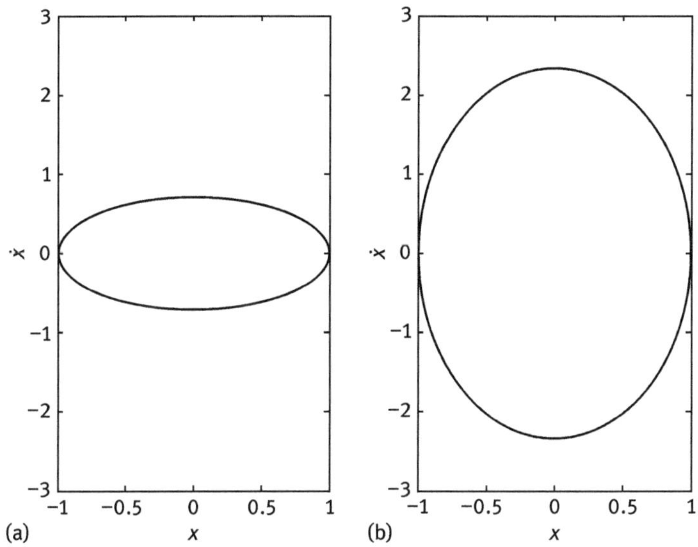

Figure 4.1: Phase plane.

In fact, control law (4.2) from arbitrary initial conditions cannot force the variables $\dot{x}$ and x toward the origin. When $0 < k < 1$, the ellipse looks like Figure 4.1(a). As shown in Figure 4.1(b), the ellipse becomes reversed when $k > 1$.

In the sense of Lyapunov, the control law (1.2) can only guarantee that the second-order system (1.1) is bounded rather than asymptotically stable. Intuitively, the combined control system may move toward the origin if the two control laws in Figure 4.1(a) and (b) are pieced together.

For the purpose of illustration, design another alternative control law (4.3) to piece the two control laws:

$$u(t) = \begin{cases} -k_1 x(t) & \text{if } x\dot{x} < 0 \\ -k_2 x(t) & \text{otherwise} \end{cases} \tag{4.3}$$

Here both k_1 and k_2 are positive.

According to the switching rule in eq. (4.3), the intersecting $\dot{x}$- and x-axes divide the phase plane in Cartesian space plane into four quadrants. $u(t) = -k_1 x(t)$ will be applied to the system in quadrants I and III of the phase plane and $u(t) = -k_2 x(t)$ will be implemented on the system in quadrants II and IV of the phase plane.

Concerning the compound control law (4.3), the system trajectory in the phase plane must spiral toward the origin. Apply the control law (4.3) to system (4.1) from the initial conditions $x(0) = 1$ and $\dot{x}(0) = 0$, where $k_1 = 0.5$ and $k_2 = 5.5$. An asymptotically stable motion is demonstrated in Figure 4.2.

From Figures 4.1 and 4.2, a stable closed-loop system has been built up by a rule of combining two control laws that cannot independently stabilize the system. The two

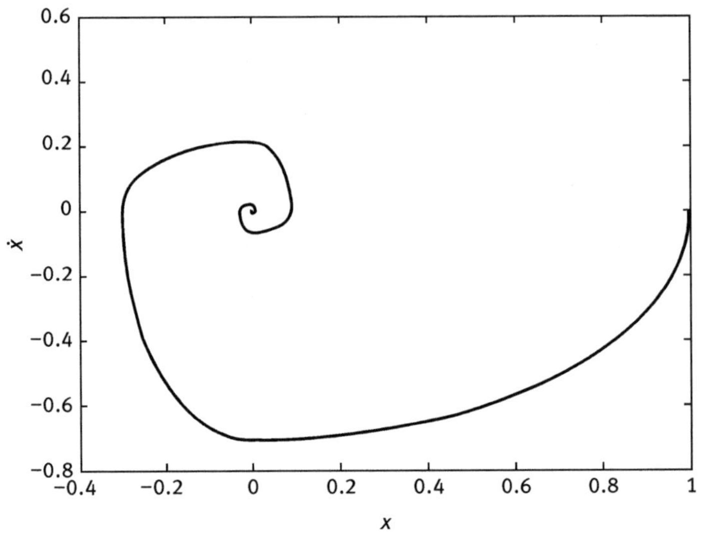

Figure 4.2: An asymptotically stable motion via the compound control law.

control laws result in two different system dynamics (structures) in the phase plane such that the design method is entitled "variable structure control."

Sliding mode control has a close relevance to variable structure control. Briefly, it is just a special case of variable structure control [3]. A characteristic that variable structure control is different from other control methods is the design of switching rules. In a large part, the switching rule (4.3) originates from intuition and analysis. To systematize the design tool of variable structure control, another variable structure control law can be given:

$$u(t) = \begin{cases} -1 & \text{if } s(x, \dot{x}) > 0 \\ 1 & \text{if } s(x, \dot{x}) < 0 \end{cases} \tag{4.4}$$

In eq. (4.4), the switching rule is determined by a function called switching function, which is defined as

$$s(x, \dot{x}) = mx + \dot{x} \tag{4.5}$$

Here $m > 0$ is scalar and $s(x, \dot{x})$ is abbreviated as s.

According to the designed switching function, the control structure is determined at any point $(x, \dot{x})$ in the phase plane. Furthermore, eq. (4.4) can be rewritten as

$$u(t) = -\text{sgn}(s) \tag{4.6}$$

In eq. (4.6), sgn() denotes a signum function, defined as

$$\mathrm{sgn}(s) = \begin{cases} 1 & \text{if } s > 0 \\ 0 & \text{if } s = 0 \\ -1 & \text{if } s < 0 \end{cases} \tag{4.7}$$

The signum function has a property of

$$s\,\mathrm{sgn}(s) = |s| \tag{4.8}$$

Apply the control law (4.6) to the second-order system (4.1). The system trajectories on either side of the following line $\mathfrak{L}_S$ will point toward the line when $m\,|\dot{x}| < 1$:

$$\mathfrak{L}_S = \left\{ (x, \dot{x}) : s(x, \dot{x}) = 0 \right\} \tag{4.9}$$

When $m = 1$, Figure 4.3 illustrates the trajectory slide along the line to the origin. Illustrated in Figure 4.3, the system trajectory repeatedly crosses the line $\mathfrak{L}_S$. Each cross indicates that the control system is switching between the two different control structures. Intuitively, such repeated crosses have a direct effect on the system performance. In fact, such switching makes the control input choppy. Displayed in Figure 4.4, the effect is the high-frequency switching of the control signal and the fluctuation of the system output.

As a result, some terms to describe sliding mode control can be drawn. The line $\mathfrak{L}_S$ is entitled the sliding surface, the trajectory sliding along the surface is termed

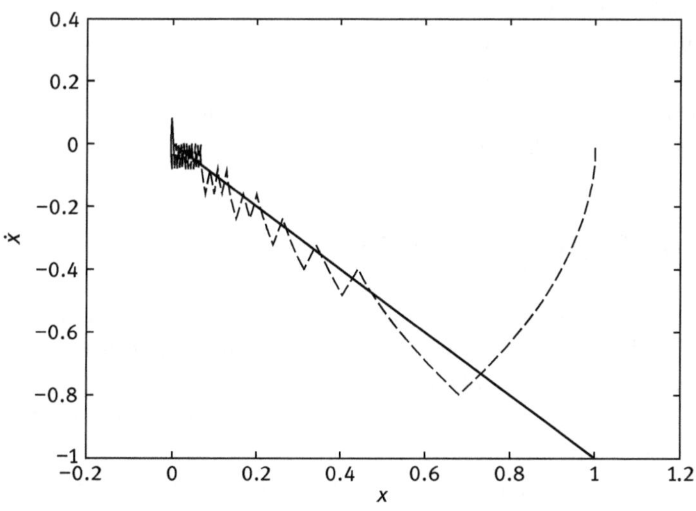

Figure 4.3: Phase portrait of a sliding motion of the system when $m = 1$.

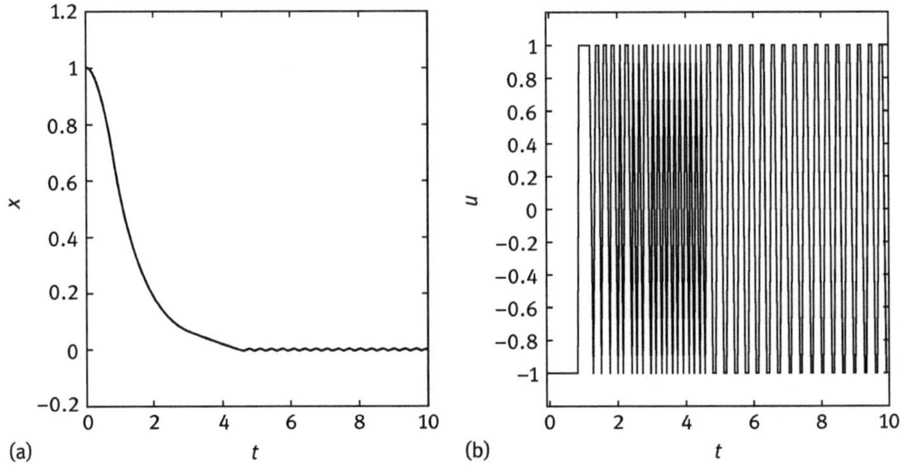

Figure 4.4: System performance when $m = 1$: (a) system output x and (b) control signal u.

the ideal sliding mode, the trajectory moving toward the surface is called the reaching mode, and the high-frequency switching of the control signal is described by chattering.

Concerning the sliding mode control, a condition can be drawn from the Lyapunov direct method, referred to as the reachability condition (4.10):

$$s\dot{s} < 0 \tag{4.10}$$

Here s is the sliding surface of a sliding mode control system.

From Figure 4.4, such a sliding mode control system suffers from the chattering phenomenon. The chattering is an inherent drawback of sliding mode control, which is also a gap between sliding mode control and real applications. In order to make the sliding-mode-based control design applicable, the equivalent-control-based sliding mode control design rises up and fills the gap.

4.2 Incremental Sliding Mode Design

Owing to the nature of strong robustness, the versatile sliding mode control method has received considerable devotion from researchers. Consider the double-pendulum-type overhead crane system (2.31). This physical structure can be divided into three components, that is, trolley, hook, and payload. Capturing the structure, the method of incremental sliding mode control is presented [4].

Recall the crane model (2.32). There are six state variables to describe the crane system. As a result, a novel type of sliding surface structure can be constructed by increasing the number of variables to produce the hierarchical structure of sliding

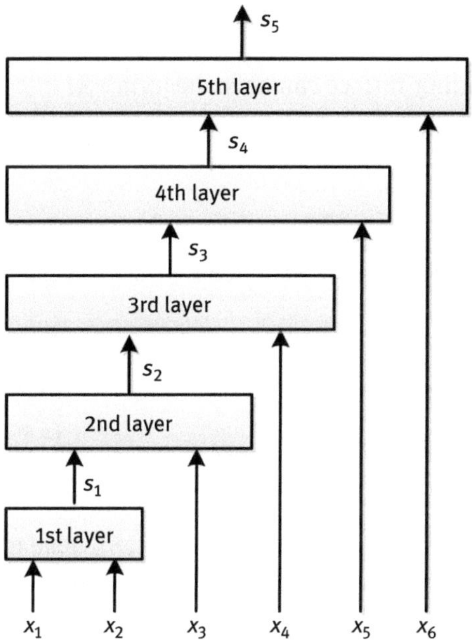

Figure 4.5: Hierarchical structure of the incremental sliding surfaces.

surfaces. The basic idea behind the type of incremental hierarchical structure [5] is as follows.

Consider the crane system (2.32). Select arbitrary two state variables to construct the first-layer sliding surface. Then, increase one state variable to construct the second-layer sliding surface with the first-layer sliding surface variable. This process proceeds till all the system state variables are included. The incremental hierarchical structure is developed by general increment of the system state variables. Since the crane model covers six state variables, the incremental sliding surfaces of the crane control system have five layers, as illustrated in Figure 4.5.

4.2.1 Control Design

The first step of the incremental sliding mode design is to construct the incremental structure of sliding surfaces. Recall the crane model (2.32). There are six state variables in the model. As mentioned earlier, arbitrary two state variables can be employed to formulate the first-layer sliding surface. Without loss of generality, the state variables x_1 and x_2 in eq. (2.32) are adopted such that the first-layer sliding surface has the form

$$s_1 = \alpha_1(x_1 - x_d) + \alpha_2 x_2 \tag{4.11}$$

where α_1 and α_2 are constants.

The second-layer sliding surface s_2 is constructed by the linear combination of the first-layer sliding surface variable s_1 and one state variable. Without loss of generality, x_3 is picked up such that the second-layer sliding surface can have the form

$$s_2 = \alpha_3 x_3 + s_1 \tag{4.12}$$

where α_3 is constant.

Similarly, other three layers of sliding surfaces can be written as

$$s_3 = \alpha_4 x_4 + s_2 \tag{4.13}$$

$$s_4 = \alpha_5 x_5 + s_3 \tag{4.14}$$

$$s_5 = \alpha_6 x_6 + s_4 \tag{4.15}$$

Here α_4, α_5 and α_6 are constants.

From the aforementioned control design, six parameters, α_i $(i = 1, \dots, 6)$, need to be designed in the hierarchical structure. In practice, the parameter-tuning of such design is a time-consuming task.

To achieve the transport control problem, the incremental hierarchical sliding mode control law [4, 5] is defined as

$$u = u_{\mathrm{eq}} + u_{\mathrm{sw}} \tag{4.16}$$

From eq. (4.16), u_{eq} is the equivalent control law and u_{sw} is the switching control law. According to control law (4.6), such kind of control design can induce the chattering phenomenon. The equivalent-control-based sliding mode control law (4.16) can make for the decrement of the chattering phenomenon.

The control law (4.16) can be deduced from the Lyapunov direct method. In order to have the stability of the last-layer sliding surface (4.15), the Lyapunov function candidate can be defined as

$$V(t) = \frac{1}{2} s_5^2 \tag{4.17}$$

Such a function in eq. (4.17) indicates that the control law (4.16) is designed to asymptotically stabilize the last-layer sliding surface. Differentiating V in eq. (4.17) with respect to time t yields

$$\frac{\mathrm{d}V}{\mathrm{d}t} = \frac{\mathrm{d}V}{\mathrm{d}s_5}\frac{\mathrm{d}s_5}{\mathrm{d}t} = s_5 \dot{s}_5 = s_5 \frac{\mathrm{d}(\alpha_6 x_6 + s_4)}{\mathrm{d}t} \tag{4.18}$$

Consider the crane model and the hierarchical sliding mode surfaces and substitute these equations into eq. (4.18). Then, eq. (4.18) can be written as

$$\begin{aligned}\frac{\mathrm{d}V}{\mathrm{d}t} &= s_5[\alpha_6(f_3 + b_3u) + \alpha_5\dot{x}_5 + \dot{s}_4] \\ &= s_5[\alpha_6(f_3 + b_3u) + \alpha_5x_6 + \alpha_4\dot{x}_4 + \dot{s}_3] \\ &= s_5[\alpha_6(f_3 + b_3u) + \alpha_5x_6 + \alpha_4(f_2 + b_2u) + \dot{s}_2] \\ &= s_5[\alpha_6(f_3 + b_3u) + \alpha_5x_6 + \alpha_4(f_2 + b_2u) + \alpha_3x_4 + \dot{s}_1] \\ &= s_5[\alpha_6(f_3 + b_3u) + \alpha_5x_6 + \alpha_4(f_2 + b_2u) + \alpha_3x_4 + \alpha_2(f_1 + b_1u) + a_1x_2]\end{aligned} \tag{4.19}$$

From eq. (4.17), $V(t) \geq 0$. In the sense of Lyapunov, $\dot{V}(t) < 0$ should exist in eq. (4.19) to have the stability of the last-layer sliding surface. To realize the purpose, the equivalent control law and the switching control law in eq. (4.16) can be defined as

$$u_{\mathrm{eq}} = -\frac{\alpha_1x_2 + a_2f_1 + \alpha_3x_4 + \alpha_4f_2 + \alpha_5x_5 + \alpha_6f_3}{\alpha_2b_1 + \alpha_4b_2 + \alpha_6b_3} \tag{4.20}$$

and

$$u_{\mathrm{sw}} = -\frac{\eta s_5 + \kappa\,\mathrm{sgn}(s_5)}{\alpha_2b_1 + \alpha_4b_2 + \alpha_6b_3} \tag{4.21}$$

Here η and κ are positive constants.

Substituting eqs. (4.20) and (4.21) into eq. (4.19) yields

$$\dot{V} = -\eta s_5^2 - \kappa|s_5| \leq 0 \tag{4.22}$$

Since the derivative of V is just the left side of the reachability condition of the last-layer sliding surface, eq. (4.22) indicates the last-layer sliding surface variable s_5 not only has the asymptotic stability, but also its sliding motion is reachable in finite time. From the control design, there are still other hierarchical sliding surfaces. But their stabilities are kept touched. To have the stabilities of all the hierarchical sliding surfaces, the following stability analysis is carried out.

4.2.2 Stability Analysis

Theorem 4.1. Consider the double-pendulum-type overhead crane system (2.32), design the incremental structure of the hierarchical sliding surfaces (4.11), (4.12), (4.13), (4.14), and (4.15), define the control law (4.16), and adopt the equivalent control law (4.20) and the switching control law (4.21). Then, all the incremental sliding surfaces are asymptotically stable if the following conditions (4.23), (4.24), and (4.25) hold true:

$$\alpha_5 = \begin{cases} \alpha_5 & \mathit{if}\; x_5s_3 \geq 0 \\ -\alpha_5 & \mathit{if}\; x_5s_3 < 0 \end{cases} \tag{4.23}$$

$$\alpha_4 = \begin{cases} \alpha_4 & if\, x_4 s_2 \geq 0 \\ -\alpha_4 & if\, x_4 s_2 < 0 \end{cases} \tag{4.24}$$

$$\alpha_3 = \begin{cases} \alpha_3 & if\, x_3 s_1 \geq 0 \\ -\alpha_3 & if\, x_3 s_1 < 0 \end{cases} \tag{4.25}$$

$$sgn(\alpha_2 \alpha_1) > 0 \tag{4.26}$$

Proof. Due to the designed hierarchical structure, the stabilities of the hierarchical sliding surfaces couple each other. In order to touch this issue, their stabilities will be considered layer by layer.

1) Stability of s_5: According to the control design, the control law (4.16) can guarantee the asymptotic stability of s_5. Moreover the last-layer sliding surface variable not only has the guaranteed stability but also its sliding mode is reachable at finite time t_r:

$$\lim_{t \to \infty} s_5 = 0$$

2) Stabilities of s_4, s_3, and s_2: Consider the fifth-layer sliding surface s_5 achieves its sliding mode at t_r. Thereafter, there is no discontinuous switching control in the control system on the time interval (t_r, ∞) and the closed-loop crane control system can be treated as an autonomous system. According to LaSalle's principle, s_5 approaches the largest invariant set in

$$\mathbb{S} = \left\{ s_5 \in \mathbb{S}_c | \frac{\mathrm{d}V}{\mathrm{d}t} = 0 \right\} \tag{4.27}$$

In eq. (4.27), the set $\mathbb{S}_c$ is a positively invariant and compact, defined as

$$\mathbb{S}_c = \left\{ s_5 \in \mathbb{R}^2 | \frac{\mathrm{d}V}{\mathrm{d}t} < c,\ c > 0 \right\} \tag{4.28}$$

Thus, there exists

$$\begin{aligned} \mathbb{S} &= \{ s_5 | s_5 = 0 \cap \dot{s}_5 = 0 \} \\ &= \{ x_6,\ s_4 | \alpha_6 x_6 + s_4 = 0 \cap \alpha_6 \dot{x}_6 + \dot{s}_4 = 0 \} \\ &= \{ x_6,\ s_4 | s_4 = -\alpha_6 x_6 = \text{const.} \cap \dot{s}_4 = -\alpha_6 \dot{x}_6 = \text{const.} \} \end{aligned} \tag{4.29}$$

Assume the two constants in eq. (4.29) are not equal to zero as $t \to \infty$. Then, s_5 would converge to a point on the phase plane by the axes x_6 and s_4 except the origin, which contradicts the fact $\lim_{t \to \infty} s_5 = 0$. Consequently, both the constants in eq. (4.29) must be equal to zero from the proof by contradiction, that is, s_4 is locally asymptotically stable. Since s_4 is locally asymptotically stable, the limit of s_4 can be drawn from

$$\lim_{t \to \infty} s_5 = \lim_{t \to \infty} (\alpha_5 x_6 + s_4) = 0 \tag{4.30}$$

Provided that the system states are not divergent, the limit of s_4 can be deduced from

$$\lim_{t\to\infty} s_4 = -\lim_{t\to\infty} \alpha_5 x_6 = \text{const.} \tag{4.31}$$

According to the given condition (4.23), eq. (4.32) can be obtained as

$$\text{sgn}(\alpha_6 x_6 s_4) \geq 0 \tag{4.32}$$

The integration of eqs. (4.31) and (4.32) indicates that the constant in eq. (4.31) is only zero rather than other constants, which means

$$\lim_{t\to\infty} x_6 = 0 \tag{4.33}$$

and

$$\lim_{t\to\infty} s_4 = 0 \tag{4.34}$$

Equation (4.34) represents that the fourth-layer sliding surface variable s_4 is locally asymptotically stable if condition (4.23) holds true.

Similarly, the stabilities of the hierarchical sliding surfaces s_3 and s_2 can be drawn in terms of the given conditions (4.24) and (4.25) hold true.

3) Stability of s_1: In eq. (4.11), x_2 is equal to the derivative of x_1. From the viewpoint of stability, $\text{sgn}(\alpha_1\alpha_2) > 0$ should be held true in eq. (4.26) such that the system trajectory can be kept in quadrants II and IV of this phase plane.

4.3 Simulation Results

To verify the feasibility and validity of such incremental sliding mode design for the anti-sway control problem of double-pendulum-type overhead cranes, some numerical simulations are carried out in this section. The control objective is to transport the payload to a desired position. Without loss of generality, the desired position x_d is located at 1 m, that is, $x_d = 1$ m from the origin. Meanwhile, the controller designed by the incremental sliding mode method needs to suppress the double-pendulum motions of the hook and payload. The results are presented in Figures. 4.6–4.16.

Concerning the double-pendulum-type crane system, its physical parameters are determined by $m = 0.5$ kg, $m_1 = 0.25$ kg, $m_2 = 0.5$ kg, $l_1 = 0.5$ m, and $l_2 = 0.5$ m, which are regarded as nominal parameters. From Chapter 2 it is known that the contribution of the low-frequency oscillation to problematic swing amplitude is particularly large for a constant $l_1 + l_2$ when the two cable lengths are approximately equal. In order to testify the controller, both of the cable lengths are set equal.

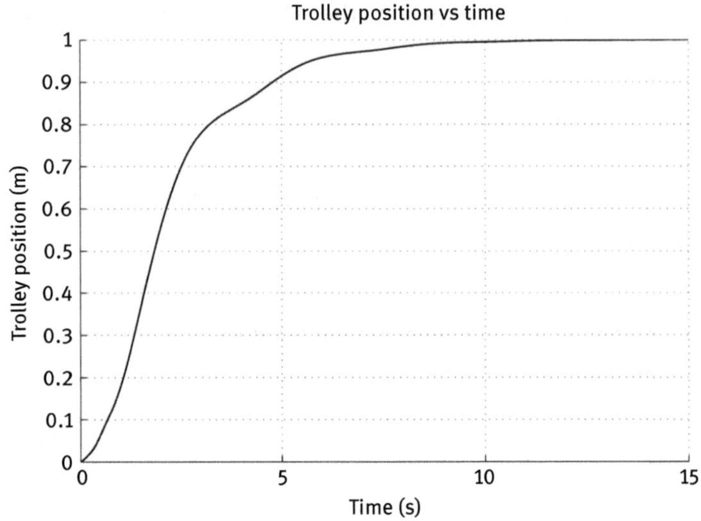

Figure 4.6: Trolley position versus time by the incremental sliding-mode-based anti-sway control.

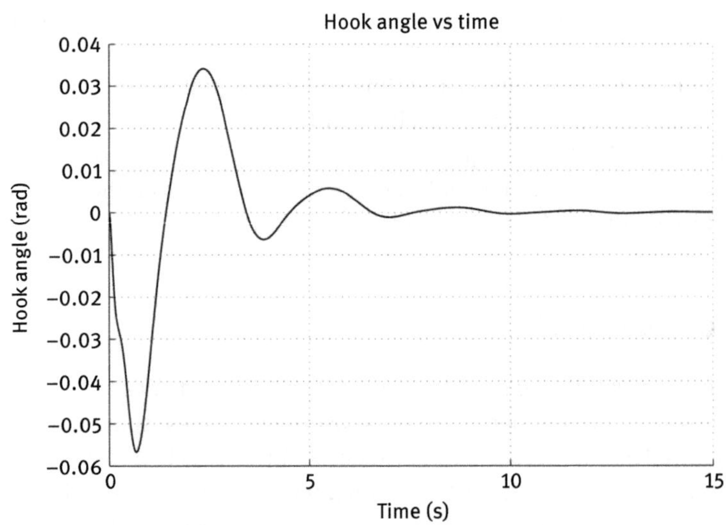

Figure 4.7: Hook angle versus time by the incremental sliding-mode-based anti-sway control.

Displayed in Figure 4.6, the incremental sliding-mode-based anti-sway controller is able to achieve the transport control objective. The trolley is stabilized at the desired position, that is, $x_d = 1$ m. The angles of payload and hook are illustrated in Figures 4.7 and 4.8. The curves indicate that the payload and hook are stabilized at the desired position. The control force applied to the trolley is displayed in Figure 4.9.

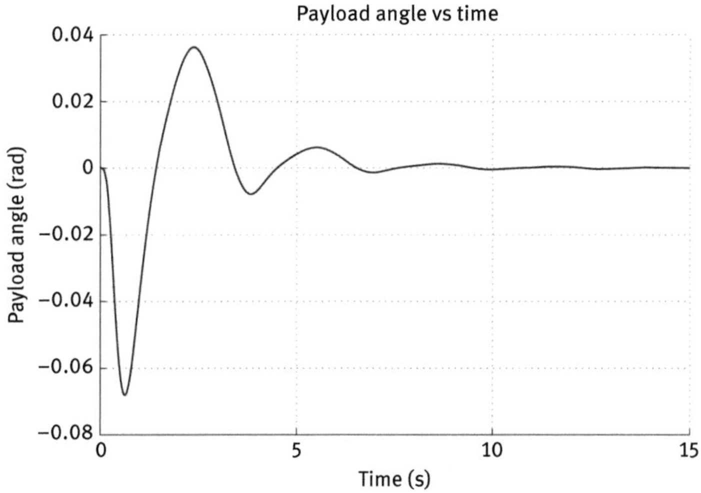

Figure 4.8: Payload angle versus time by the incremental sliding-mode-based anti-sway control.

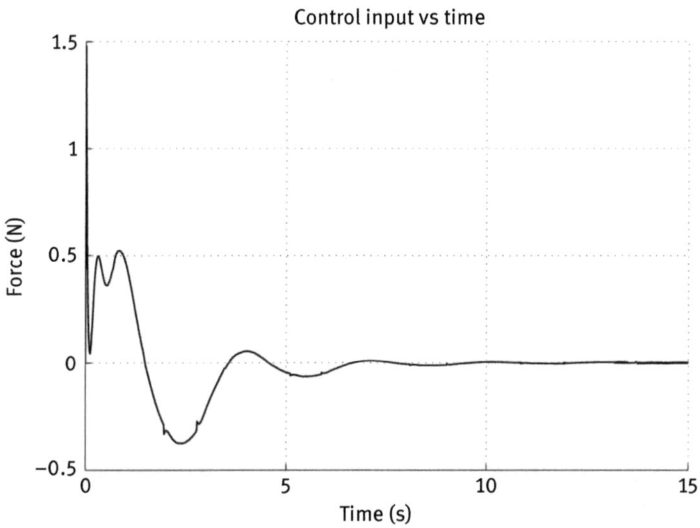

Figure 4.9: Control input versus time by the incremental sliding-mode-based anti-sway control.

The angular velocities of payload and hook are shown in Figures 4.10 and 4.11. As we have proven, all the sliding surfaces in the designed hierarchical structure are asymptotically stable. These sliding surfaces are illustrated in Figures 4.12 and 4.16.

From Figure 4.6, the trolley arrives at the desired position in about 10 s. Meanwhile, the angles of payload and hook can be stabilized at their download positions.

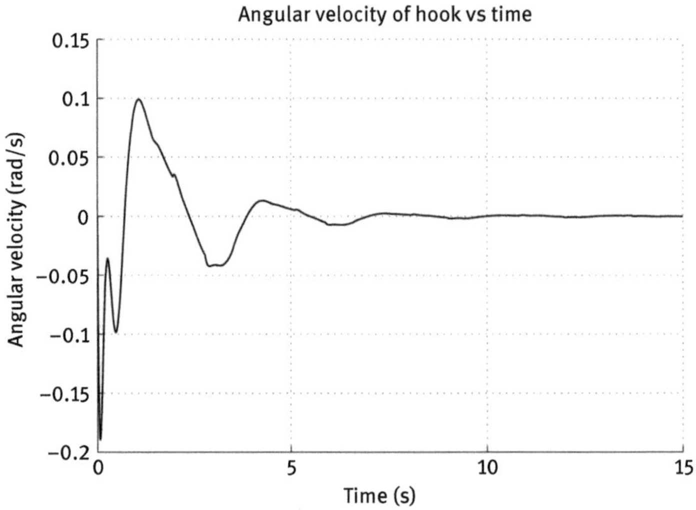

Figure 4.10: Angular velocity of hook versus time by the incremental sliding-mode-based anti-sway control.

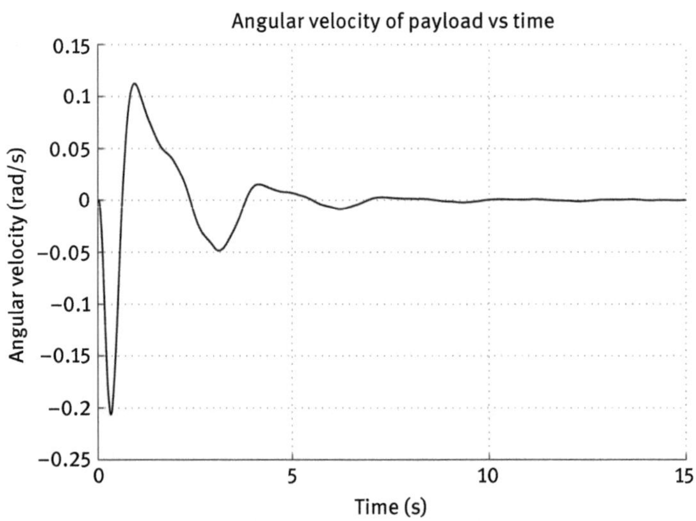

Figure 4.11: Angular velocity of payload versus time by the incremental sliding-mode-based anti-sway control.

Further, the incremental sliding mode controller is completely based on the crane model. In practice, such kind of control design is subjected to the crane model. If the model is not accurate enough, the control performance may not be as good as expected.

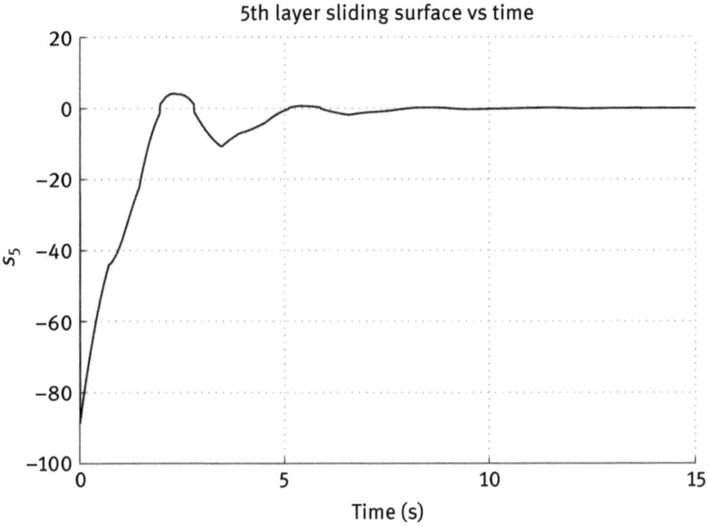

Figure 4.12: The last-layer sliding surface s_5 versus time by the incremental sliding-mode-based anti-sway control.

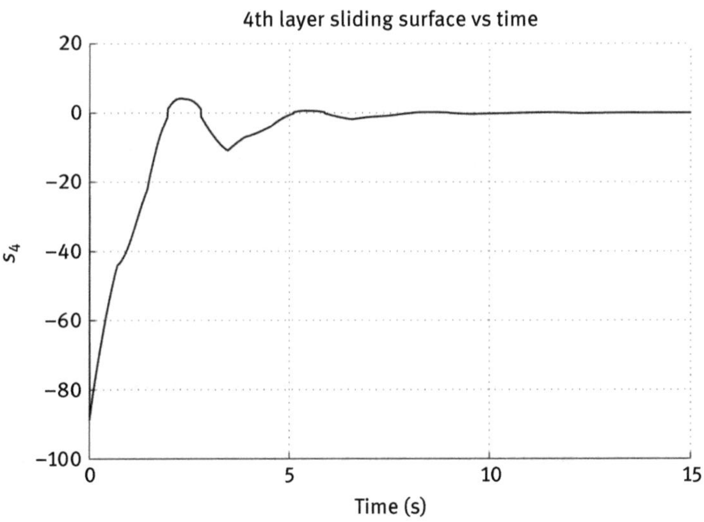

Figure 4.13: The fourth-layer sliding surface s_4 versus time by the incremental sliding-mode-based anti-sway control.

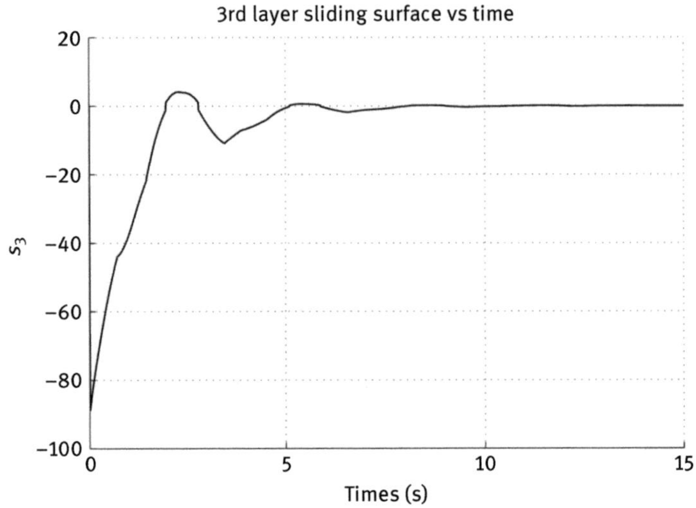

Figure 4.14: The third-layer sliding surface s_3 versus time by the incremental sliding-mode-based anti-sway control.

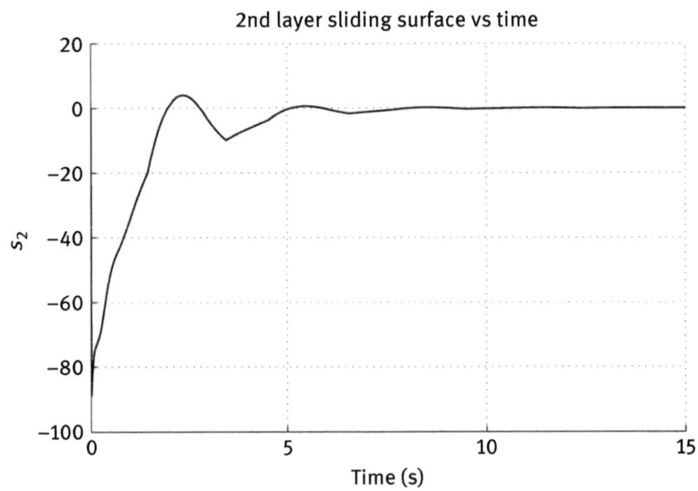

Figure 4.15: The second-layer sliding surface s_3 versus time by the incremental sliding-mode-based anti-sway control.

4.4 Hierarchical Sliding Mode Design

4.4.1 Control Design

Tuan and Lee [6] investigated the transport control problem by the hierarchical sliding mode design for double-pendulum-type overhead cranes. Their design captures the

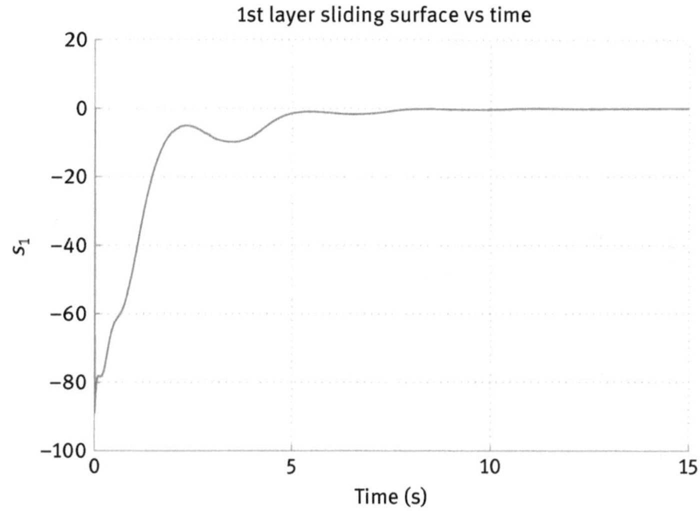

Figure 4.16: The first-layer sliding surface s_3 versus time by the incremental sliding-mode-based anti-sway control.

physical structure of double-pendulum-type overhead cranes, that is, these kinds of cranes consist of three subsystems: trolley, hook, and payload. Recall the mathematic model of double-pendulum-type overhead cranes (2.32). There are six state variables in eq. (2.32). Each subsystem can be described by two state variables.

From Ref. [6], the sliding surface vector **s** can be defined in order to construct the hierarchical structure of sliding surfaces, formulated as

$$\mathbf{s} = \begin{bmatrix} s_1 \\ s_2 \\ s_3 \end{bmatrix} = \begin{bmatrix} \lambda_1(x_1 - x_d) + x_2 \\ \lambda_2 x_3 + x_4 \\ \lambda_3 x_5 + x_6 \end{bmatrix} \tag{4.35}$$

Here λ_1, λ_2, and λ_3 are constants. Each element in eq. (4.35) depicts a subsystem sliding surface. The three subsystem sliding surfaces are aggregated together to build up the first layer of the hierarchical structure, as illustrated in Figure 4.17. Based on that structure, the second-layer sliding surface is defined as

$$S = \boldsymbol{\mu}^T \mathbf{s} = \mu_1 s_1 + \mu_2 s_2 + \mu_3 s_3 \tag{4.36}$$

where $\boldsymbol{\mu} = [\mu_1\ \mu_2\ \mu_3]^T$ and μ_1, μ_2, and μ_3 are constants.

At the first glance, the hierarchical structure defined by eq. (4.36) looks like a simple combination of the system state variables. In fact, such a combination raises some novel control problems, that is, to design the hierarchical sliding mode control law and to guarantee the stability analysis of each sliding surface.

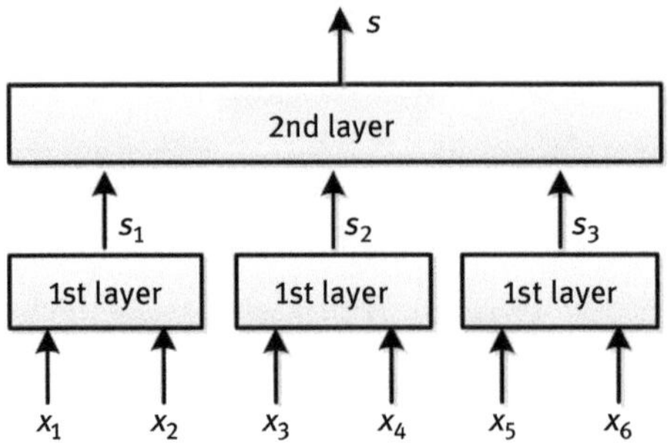

Figure 4.17: Hierarchical structure of the sliding surfaces.

To achieve the transport control problem, the hierarchical sliding mode control law [6] is defined as

$$u = u_{\text{eq}} + u_{\text{sw}} \tag{4.37}$$

In eq. (4.37), u_{eq} is the equivalent control law and u_{sw} is the switching control law, where the equivalent control law is defined as

$$u_{\text{eq}} = u_{\text{eq1}} + u_{\text{eq2}} + u_{\text{eq3}} \tag{4.38}$$

where u_{eq1}, u_{eq2}, and u_{eq3} delegate the equivalent control law of the three subsystems.

When the state trajectories of the three subsystems reach their own subsystem sliding surfaces at the first layer, the equivalent control laws given in Ref. [6] can be deduced from

$$\dot{\mathbf{s}} = 0 \tag{4.39}$$

These equivalent control laws can be formulated as

$$\begin{aligned} u_{\text{eq1}} &= -\frac{1}{b_1(\mathbf{x})}[\lambda_1 x_2 + f_1(\mathbf{x})] \\ u_{\text{eq2}} &= -\frac{1}{b_2(\mathbf{x})}[\lambda_2 x_4 + f_2(\mathbf{x})] \\ u_{\text{eq3}} &= -\frac{1}{b_3(\mathbf{x})}[\lambda_3 x_6 + f_3(\mathbf{x})] \end{aligned} \tag{4.40}$$

Further, the switching control law u_{sw} has a switching action with sufficiently high frequency to retain all state trajectories moving on the second-layer sliding surface $S = 0$. From eq. (4.36), S denotes the switching line that all system states are attracted to and retained on. To have the reachability of sliding mode at the aggregated sliding surface S, a Lyapunov function candidate can be selected as

$$V = \frac{1}{2}S^2 \tag{4.41}$$

Differentiating V with respect to time t yields

$$\dot{V} = S\dot{S} \tag{4.42}$$

Differentiating S with respect to time t in eq. (4.36) and substituting the derivative of S into eq. (4.42) yield

$$\dot{V} = S(\mu_1\dot{s}_1 + \mu_2\dot{s}_2 + \mu_3\dot{s}_3) \tag{4.43}$$

Consider the crane model (2.32) and substitute eqs. (4.38) and (4.40) into eq. (4.43). Then, eq. (4.43) becomes

$$\begin{aligned}\dot{V} &= S\{\mu_1 b_1(\mathbf{x})[u_{eq2} + u_{eq3} + u_{sw}] + \mu_2 b_2(\mathbf{x})[u_{eq1} + u_{eq3} + u_{sw}] \\ &\quad + \mu_3 b_3(\mathbf{x})[u_{eq1} + u_{eq2} + u_{sw}]\} \\ &= S\{[\mu_1 b_1(\mathbf{x}) + \mu_2 b_2(\mathbf{x}) + \mu_3 b_3(\mathbf{x})]u_{sw} + \mu_1 b_1(\mathbf{x})(u_{eq2} + u_{eq3}) \\ &\quad + \mu_2 b_2(\mathbf{x})(u_{eq1} + u_{eq3}) + \mu_3 b_3(\mathbf{x})(u_{eq1} + u_{eq2})\}\end{aligned} \tag{4.44}$$

In eq. (4.41), $V(t) \geq 0$. In the sense of Lyapunov, the derivative of S should be larger than zero in eq. (4.44) to make the crane control system asymptotically stable. For this purpose, the switching control u_{sw} can be defined as

$$\begin{aligned}u_{sw} &= -\frac{\mu_1 b_1(\mathbf{x})(u_{eq2} + u_{eq3}) + \mu_2 b_2(\mathbf{x})(u_{eq1} + u_{eq3}) + \mu_3 b_3(\mathbf{x})(u_{eq1} + u_{eq2})}{\mu_1 b_1(\mathbf{x}) + \mu_2 b_2(\mathbf{x}) + \mu_3 b_3(\mathbf{x})} \\ &\quad - \frac{\kappa S + \eta \operatorname{sgn}(S)}{\mu_1 b_1(\mathbf{x}) + \mu_2 b_2(\mathbf{x}) + \mu_3 b_3(\mathbf{x})}\end{aligned} \tag{4.45}$$

where κ and η are positive constants.

Substitute eqs. (4.38), (4.40), and (4.45) into eq. (4.37). The control law based on the hierarchical structure of sliding surfaces in Figure 4.17 [7] has the form

$$\begin{aligned}u &= -\frac{\mu_1 b_1(\mathbf{x})u_{eq1} + \mu_2 b_2(\mathbf{x})u_{eq2} + \mu_3 b_3(\mathbf{x})u_{eq3}}{\mu_1 b_1(\mathbf{x}) + \mu_2 b_2(\mathbf{x}) + \mu_3 b_3(\mathbf{x})} \\ &\quad - \frac{\kappa S + \eta \operatorname{sgn}(S)}{\mu_1 b_1(\mathbf{x}) + \mu_2 b_2(\mathbf{x}) + \mu_3 b_3(\mathbf{x})}\end{aligned} \tag{4.46}$$

The control design in the sense of Lyapunov can guarantee that the sliding surface S located at the top layer is asymptotically stable. How, the stabilities of other subsystem sliding surfaces in the hierarchical structure are not touched because they have no guaranteed stabilities from the control design.

4.4.2 Stability Analysis

Theorem 4.2. Consider the double-pendulum-type overhead crane system (2.32), design the hierarchical sliding surfaces (4.35) and (4.36), adopt the hierarchical sliding mode control law (4.37), define the equivalent control laws of the three subsystems as eq. (4.40), and define the switching control law as eq. (4.45). Then, the subsystem sliding surfaces located at the first layer s_1, s_2, and s_3 in eq. (4.35) are asymptotically stable.

Proof. Substitute the switching control law (4.45) into the derivative of V (4.44). Then, we can obtain $\dot{V}(t) < 0$. Concerning $V(t) \geq 0$ in eq. (4.41), the fact indicates that the reachability condition of sliding mode is satisfied on the second-layer sliding surface and that the sliding mode of the second-layer sliding surface is reachable in the finite time t_f.

Consequently, the time domain can be divided into two parts by the time point t_f. On the time interval $[0, t_f]$, the system state trajectory moves toward the second-layer sliding surface S. On the subsequent time interval (t_f, ∞), the system state trajectory keeps sliding on the second-layer sliding surface and converges to the origin.

To check the stabilities of s_1 and s_2 in $[0, t_f]$, define a truncated function as

$$f_{\mathrm{tf}}(t) = \begin{cases} 1 & t \leq t_{\mathrm{f}} \\ 0 & t > t_{\mathrm{f}} \end{cases} \tag{4.47}$$

Consequently, there exists

$$s_{i_{\mathrm{tf}}} = s_i \cdot f_{\mathrm{tf}}(t) \tag{4.48}$$

and

$$\dot{s}_{i_{\mathrm{tf}}} = \dot{s}_i \cdot f_{\mathrm{tf}}(t) \tag{4.49}$$

where $i = 1, 2$, and 3.

Equations (4.48) and (4.49) represent the first-layer sliding surface variables s_1, s_2, and s_3 are bounded on the closed interval $[0, t_f]$ and all of them can converge to the second-layer sliding surface when the sliding mode is reached at t_f. On the subsequent open interval (t_f, ∞), the system trajectory maintains on the second-layer sliding surface S and the closed-loop crane control system becomes autonomous.

In order to analyze the stabilities of the three first-layer sliding surfaces, a positively invariant and compact set is defined as

$$\mathbb{S}_c = \left\{ S \in \mathbb{R}^2 | \dot{V} \leq 0 \right\} \tag{4.50}$$

Further, define

$$\mathbb{S} = \left\{ S \in \mathbb{S} | \dot{V} = 0 \right\} \tag{4.51}$$

According to LaSalle's principle, the system trajectory tends to the largest invariant set inside $\mathbb{S}$ as $t \to \infty$ in $\mathbb{S} \cap \mathbb{S}_c$. To calculate the largest invariant set, $\mathbb{S} \cap \mathbb{S}_c$ can be rewritten as

$$\mathbb{S}_c \cap \mathbb{S} = \mathbb{S}_c \cap \left\{ S = 0 \quad \dot{S} = 0 \right\} \tag{4.52}$$

Then, eq. (4.53) can be drawn from eq. (4.52):

$$\mathbb{S}_c \cap \mathbb{S} = \mathbb{S}_c \cap \{\mu_1 s_1 + \mu_2 s_2 + \mu_3 s_3 = 0 \quad \mu_1 \dot{s}_1 + \mu_2 \dot{s}_2 + \mu_3 \dot{s}_3 = 0\} \tag{4.53}$$

Since the second-layer sliding surface variable S is asymptotically stable, the fact that indicates the phase trajectory of the second-layer sliding surface S will enter the neighborhood of the three-dimensional coordinate origin constructed by the axes s_1, s_2, and s_3 as $t \to \infty$. Subsequently, each subsystem states will move on the first-layer sliding surface and toward its subsystem coordinate origin constructed by the subsystem state variables such that eq. (4.54) can be drawn:

$$\begin{cases} s_1 = \dot{s}_1 = 0 \\ s_2 = \dot{s}_2 = 0 \\ s_3 = \dot{s}_3 = 0 \end{cases} \tag{4.54}$$

As a result, the largest invariant set inside $\mathbb{S} \cap \mathbb{S}_c$ is just the three-dimensional coordinate origin constructed by the sliding surface variables s_1, s_2, and s_3 rather than other points. In light of LaSalle's principle, all the first-layer sliding surfaces s_1, s_2, and s_3 are of asymptotic stability.

4.5 Simulation Results

To verify the feasibility and validity of such incremental sliding mode design for the anti-sway control problem of double-pendulum-type overhead cranes, some numerical simulations are carried out in this section. The control objective is to transport the payload to a desired position. Without loss of generality, the desired position x_d is located at 1 m, that is, $x_d = 1$ m from the origin. Meanwhile, the controller designed by the incremental sliding mode method needs to suppress the double-pendulum motions of the hook and payload. The results are presented in Figures 4.18–4.27.

Concerning the double-pendulum-type crane system, its physical parameters are determined as $m = 0.5$ kg, $m_1 = 0.25$ kg, $m_2 = 0.5$ kg, $l_1 = 0.5$ m, and $l_2 = 0.5$ m, which

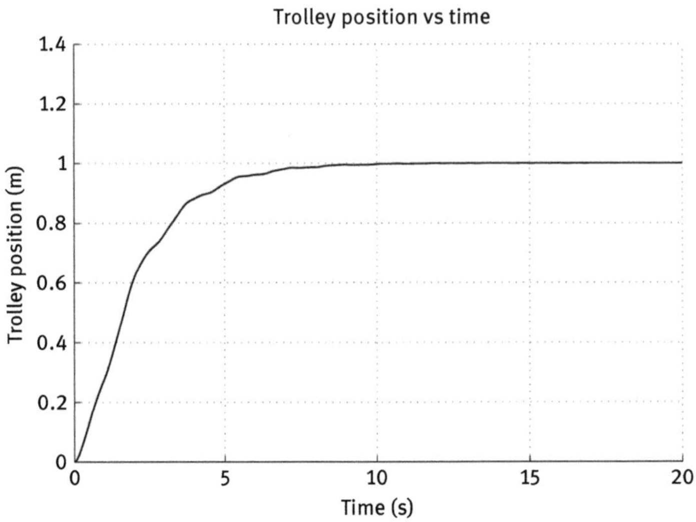

Figure 4.18: Trolley position versus time by the hierarchical sliding mode-based anti-sway control.

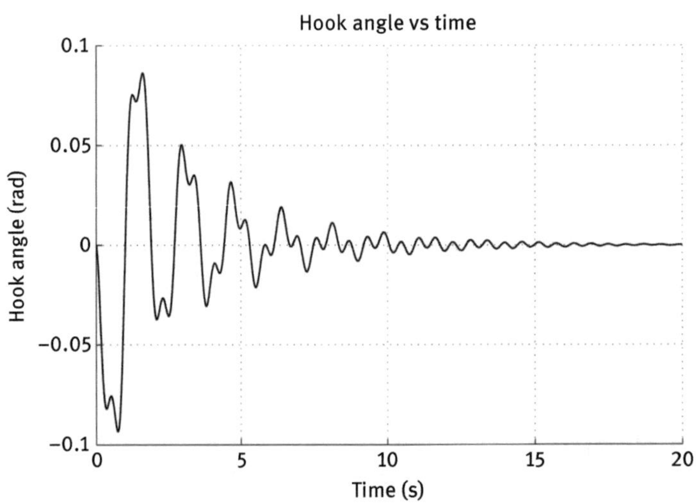

Figure 4.19: Hook angle versus time by the hierarchical sliding mode-based anti-sway control.

are regarded as nominal parameters. In order to testify the controller, both of the cable lengths are set equal. The controller parameters are determined by $\kappa = 0.5$, $\eta = 0.02$, $\lambda_1 = 10$, $\lambda_2 = 0.2$, $\lambda_3 = -15$, $\mu_1 = 4.5$, $\mu_2 = 0.05$, and $\mu_3 = -0.25$.

As displayed in Figure 4.18, the hierarchical sliding-mode-based anti-sway controller is able to achieve the transport control objective. The trolley is stabilized at the

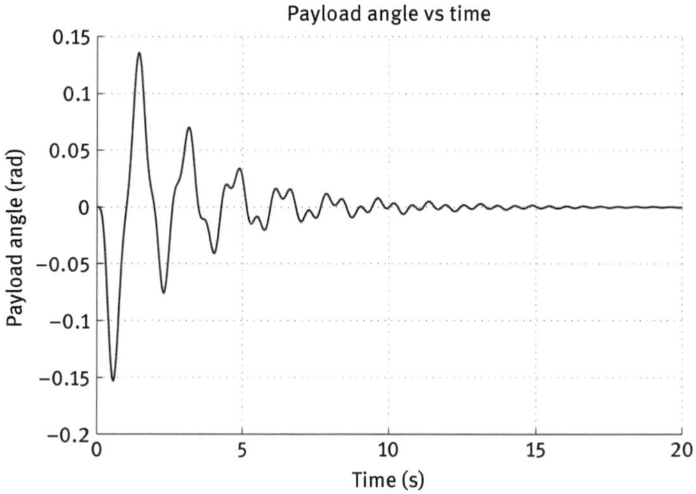

Figure 4.20: Payload angle versus time by the hierarchical sliding-mode-based anti-sway control.

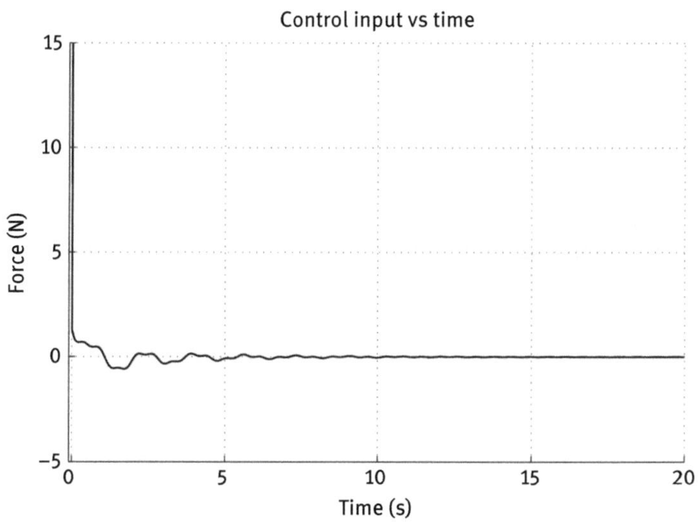

Figure 4.21: Control input versus time by the hierarchical sliding-mode-based anti-sway control.

desired position, that is, $x_d = 1$ m. The angles of payload and hook are illustrated in Figures 4.19 and 4.20. The curves indicate that the payload and hook are stabilized at the desired position. The control force applied to the trolley is displayed in Figure 4.21.

The angular velocities of payload and hook are shown in Figures 4.22 and 4.23. The curves of the angular velocities indicate that the payload and hook slightly oscillate

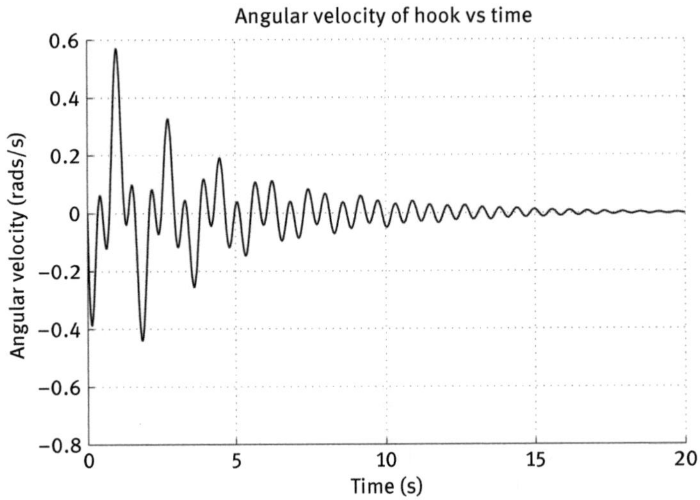

Figure 4.22: Angular velocity of hook versus time by the hierarchical sliding-mode-based anti-sway control.

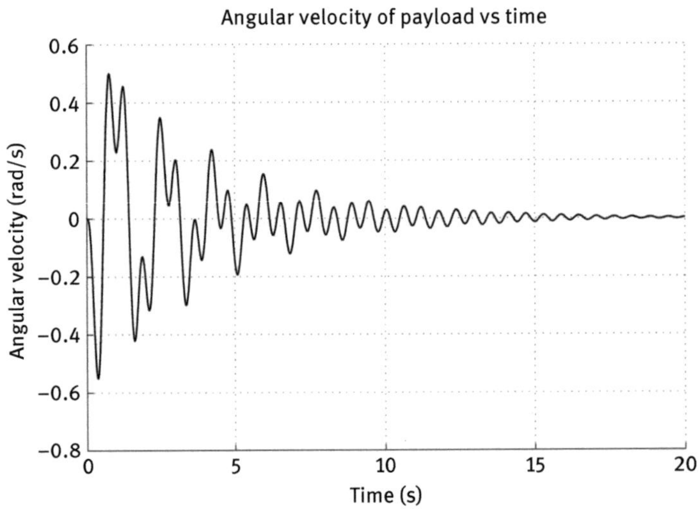

Figure 4.23: Angular velocity of payload versus time by the hierarchical sliding-mode-based anti-sway control.

when the trolley arrives at the desired position. But the oscillations are decreasing and the control objective has been achieved.

As we have proven, all the sliding surfaces in the designed hierarchical structure are asymptotically stable. These sliding surfaces are illustrated in Figures 4.24

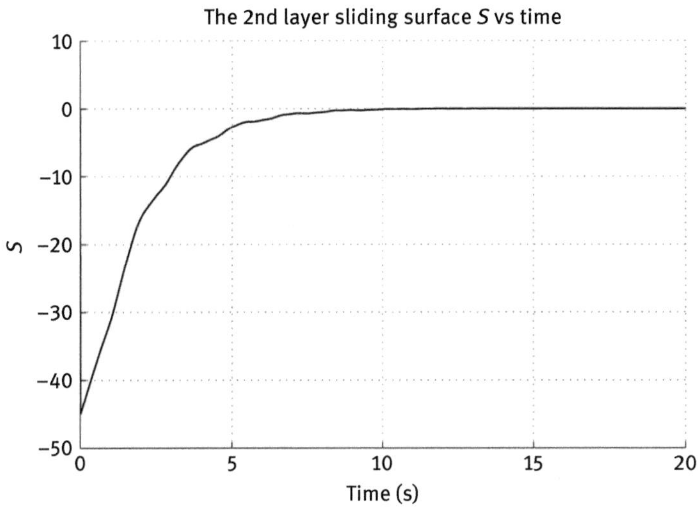

Figure 4.24: The second-layer sliding surface S versus time by the hierarchical sliding-mode-based anti-sway control.

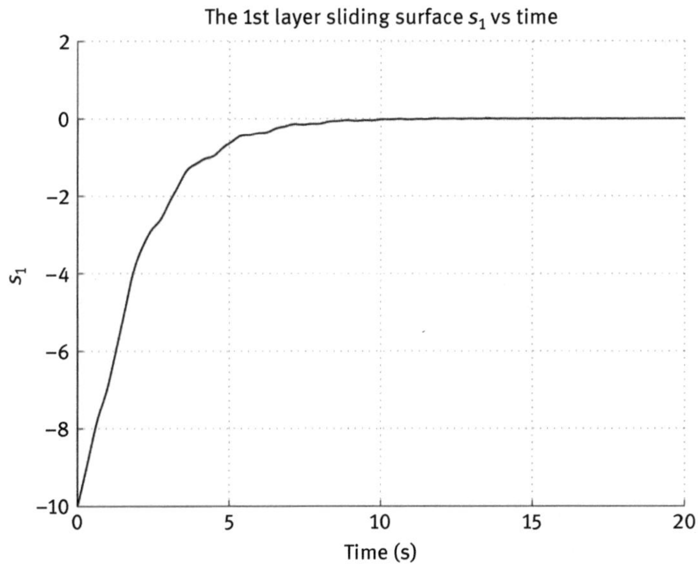

Figure 4.25: The first-layer sliding surface s_1 versus time by the hierarchical sliding-mode-based anti-sway control.

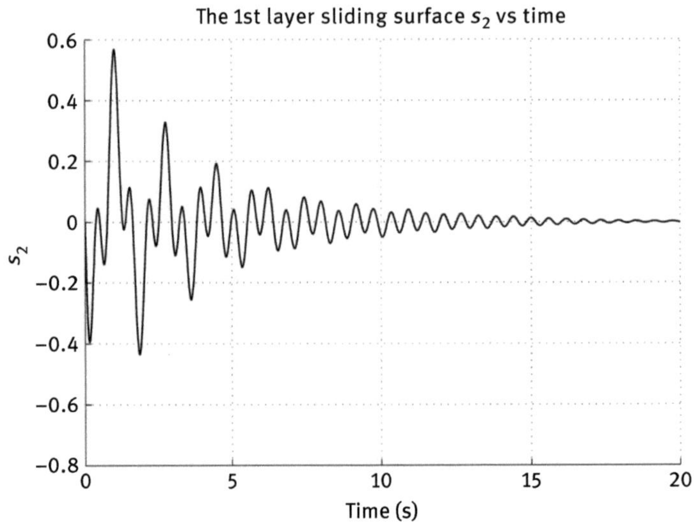

Figure 4.26: The first-layer sliding surface s_2 versus time by the hierarchical sliding-mode-based anti-sway control.

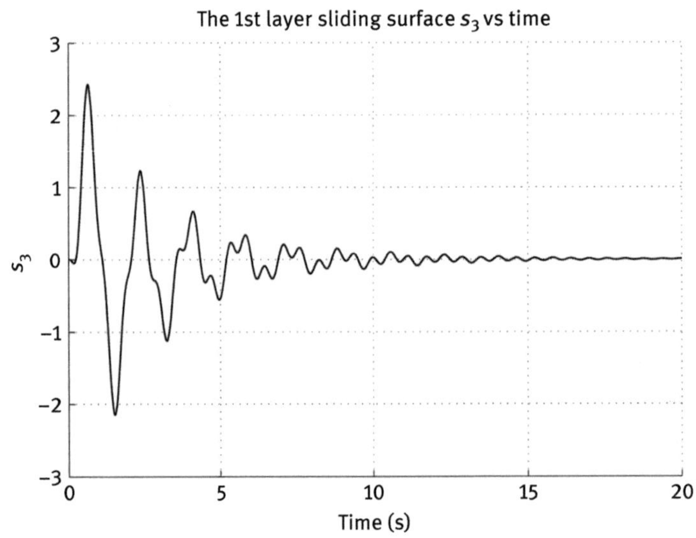

Figure 4.27: The first-layer sliding surface s_3 versus time by the hierarchical sliding-mode-based anti-sway control.

and 4.27. Pointed out by Theorem 4.2, the sliding modes of the surfaces are hierarchical, where the sliding mode of the second-layer sliding surface can be first arrived. Thereafter, the sliding mode of each subsystem converges to the origin along its own subsystem sliding surface at the first layer.

4.6 Source Codes

Figures 4.1–4.4, 4.6–4.16, and 4.18–4.27 can be numerically obtained by Matlab software [8]. The source codes are listed as follows:

Matlab codes to plot Figure 4.1

```
sol=dsolve('Dx1=x2','Dx2=-0.5*x1','x1(0)=1',
    'x2(0)=0','t');
tval=(0:0.1:10)'; % column vector with t-values
yval=double(subs([sol.x1,sol.x2],'t',tval));
sol=dsolve('Dx3=x4','Dx4=-5.5*x3','x3(0)=1',
    'x4(0)=0','t');
ttval=(0:0.1:10)'; % column vector with t-values
yyval=double(subs([sol.x3,sol.x4],'t',ttval));
subplot(1,2,1),plot(yval(:,1),yval(:,2),'k')
    % plot col.2 of yval vs. col.1 of yval
axis([-1,1,-1, ]); xlabel('x');
latexStr=['$$\dot x$$'] ;
lgh=ylabel(latexStr);
set(lgh,'interpreter','latex');
subplot(1,2,2),plot(yyval(:,1),yyval(:,2),'k')
    % plot col.2 of yval vs. col.1 of yval
axis([-1,1,-3,3]); xlabel('x');
latexStr=['$$\dot x$$'] ;
lgh=ylabel(latexStr);
set(lgh,'interpreter','latex')
```

Matlab codes to plot Figure 4.2

```
y1(1)=1;y2(1)=0;
dt=0.001;
```

```
for i=1:1:30000
     if((y1(i)*y2(i))<0)
         k(i)=0.5;
     else
         k(i)=5.5;
     end
     y1(i+1)=y2(i)*dt+y1(i);
     y2(i+1)=y2(i)-dt*k(i)*y1(i);
end
plot(y1,y2,'k');
xlabel('x');
latexStr=['$$\dot x$$'] ;
lgh=ylabel(latexStr);
set(lgh,'interpreter','latex');
```

Matlab codes to plot Figures 4.3 and 4.4

```
function signSMC
clear all;
close all;
% sliding surface
m=1;
% initial conditions
x=1;
dx=0;
dt=0.08; %sample time
t=0;
%
x_1=x;
dx_1=dx;

xdata=x;
dxdata=dx;
sdata=[1];
tdata=t;
udata=[-1];
```

```
for i=1:8:1000 %time
   x=x_1;
   dx=dx_1;
   s=m*x+dx;
   %control input
   if (s>0)
       u=-1;
   else if (s<0)
       u=1;
       else
           u=0;
       end
   end
   dx_1=dx+u*dt;
   x_1=x+dx*dt+0.5*u*dt^2;
   t=t+dt;
   xdata=[xdata,x_1];
   dxdata=[dxdata,dx_1];
   sdata=[sdata,s];
   tdata=[tdata,t];
   udata=[udata,u];
end

%plot
figure(1) %phase portrait of m=1
plot(xdata,dxdata);
xlabel('x');
latexStr=['$$\dot x$$'] ;
lgh=ylabel(latexStr);
set(lgh,'interpreter','latex');hold on;
plot([0,1],[0,-1],[0,1],[0,-1],'k-.');
hold on;

figure(2)%% control signal
subplot(1,2,1),plot(tdata,xdata);
```

```
xlabel('t');ylabel('x');
subplot(1,2,2),plot(tdata,udata,'k');
xlabel('t');ylabel('u');axis([-0.1,5,-1.2,1.2]);
```

Matlab codes to plot Figures 4.6-4.16

```
function Sliding_mode
clear all;
close all;

len1=0.5; %1/2á112ÎÊý
len2=0.5;
mass=0.5;
mass1=0.25;
mass2=0.5;
g=9.81;

s=0;
v=0;
surf1=0;
surf2=0;
surf3=0;
surf4=0;
surf5=0;
th1=0;
th2=0;
w1=0;
w2=0;
f=0;
t=0;
s_1=s;
v_1=v;
th_1=th1;
th_2=th2;
w_1=w1;
```

```
w_2=w2;
dt=0.01;
xend=1;

sdata=s;
vdata=v;
th1data=th1;
th2data=th2;
w1data=w1;
w2data=w2;
fdata=f;
tdata=t;
surf1data=surf1;
surf2data=surf2;
surf3data=surf3;
surf4data=surf4;
surf5data=surf5;

for i=1:1:1500
s=s_1;
v=v_1;
th1=th_1;
th2=th_2;
w1=w_1;
w2=w_2;

M=[mass+mass1+mass2 (mass1+mass2)*len1*cos(th1)
       mass2*len2*cos(th1);...
  (mass1+mass2)*len1*cos(th1) (mass1+mass2)*len1^2
       mass2*len1*len2* cos(th1-th2);...
  mass2*len2*cos(th2) mass2*len1*len2*cos(th1-th2)
       mass2*len2^2];
C=[0 -1*(mass1+mass2)*len1*w1*sin(th1) -1*
       mass2*len2*w2*sin(th2);...
```

```
  0 0 mass2*len1*len2*w1*sin(th1-th2);...
  0 -1*mass2*len1*len2*w1*sin(th1-th2) 0];
G=[0;(mass1+mass2)*g*len1*sin(th1);mass2*g*len2*sin(th2)];

A=[0 1 0 0 0 0;0 0 ((mass1+mass2)*g+len2*mass2*w1*w2)
     /mass 0 ...
  -1*len2*mass2*w1*w2/mass 0;0 0 0 1 0 0;0 0 ...
  (-1*(mass+mass1)*(mass1+mass2)*g-len2*mass2*w1*w2)
     /(len1*mass*mass1)...
  0 (mass*mass2*g+len2*mass2*w1*w2)/(mass*mass1*len1)
     0;0 0 0 0 0 1;...
  0 0 ((mass1+mass2)*g+mass2*len2*w1*w2)
     /(mass1*len2) 0 ...
  (-1*mass2*len2*w1*w2-(mass1+mass2)*g)
     /(mass1*len2) 0];
B=[0;1/mass;0;-1/(mass*len1);0;0];
P=[-2,-4,-8,-10,-16,-20];
alpha=place(A,B,P);

surf1=alpha(1)*(-xend+s)+alpha(2)*v;
surf2=alpha(3)*th1*sign(th1*surf1)+surf1;
surf3=alpha(4)*w1*sign(w1*surf2)+surf2;
surf4=alpha(5)*th2*sign(th2*surf3)+surf3;
surf5=alpha(6)*w2+surf4;

f=controller(A,B,M,C,G,xend,s,v,th1,th2,w1,w2 );
T=[f;0;0];
x=inv(M)*(T-G-C*[v;w1;w2]);
y=[v;w1;w2]+dt*x;
v_1=y(1);
w_1=y(2);
w_2=y(3);
z=[s;th1;th2]+0.5*dt^2*x+dt*[v;w1;w2];
s_1=z(1);
```

```
th_1=z(2);
th_2=z(3);
t=t+dt;
   tdata=[tdata,t];
   fdata=[fdata,f];
   surf1data = [surf1data,surf1];
   surf2data = [surf2data,surf2];
   surf3data = [surf3data,surf3];
   surf4data = [surf4data,surf4];
   surf5data = [surf5data,surf5];
   vdata = [vdata,v_1];
th1data=[th1data,th_1];
th2data=[th2data,th_2];
w1data=[w1data,w_1];
w2data=[w2data,w_2];
sdata=[sdata,s_1];
end

h1=figure(1);
set(h1,'Position',[100,200,500,300],...
      'Menu','none',...
      'NumberTitle','off',...
      'Name','Trolley position'...
);
axes1=axes('Parent',h1);
title(axes1,'Trolley position vs time','FontSize',16);
xlabel(axes1,'Time(s)');
ylabel(axes1,'Trolley position(m)');
hold(axes1,'all');
plot(tdata,sdata);
grid on;

h2=figure(2);
set(h2,'Position',[800,200,500,300],...
```

```
      'Menu','none',...
      'NumberTitle','off',...
      'Name','Hook angle'...
);
axes2=axes('Parent',h2);
title(axes2,'Hook angle vs time','FontSize',16);
xlabel(axes2,'Time(s)');
ylabel(axes2,'Hook angle(rad)');
hold(axes2,'all');
plot(tdata,th1data,'b');
grid on;

h2=figure(3);
set(h2,'Position',[800,200,500,300],...
      'Menu','none',...
      'NumberTitle','off',...
      'Name','Payload angle'...
);
axes2=axes('Parent',h2);
title(axes2,'Payload angle vs time','FontSize',16);
xlabel(axes2,'Time(s)');
ylabel(axes2,'Payload angle(rad)');
hold(axes2,'all');
plot(tdata,th2data,'b');
grid on;

h2=figure(4);
set(h2,'Position',[800,200,500,300],...
      'Menu','none',...
      'NumberTitle','off',...
      'Name','Control input'...
);
axes2=axes('Parent',h2);
title(axes2,'Control input vs time','FontSize',16);
```

```
xlabel(axes2,'Time(s)');
ylabel(axes2,'Force(N)');
hold(axes2,'all');
plot(tdata,fdata,'b');
grid on;

h2=figure(5);
set(h2,'Position',[800,200,500,300],...
      'Menu','none',...
      'NumberTitle','off',...
      'Name','Angular velocity of hook'...
);
axes2=axes('Parent',h2);
title(axes2,'Angular velocity of hook vs time',
   'FontSize',16);
xlabel(axes2,'Time(s)');
ylabel(axes2,'Angular velocity (rad/s)');
hold(axes2,'all');
plot(tdata,w1data,'b');
grid on;

h2=figure(6);
set(h2,'Position',[800,200,500,300],...
      'Menu','none',...
      'NumberTitle','off',...
      'Name','Angular velocity of payload'...
);
axes2=axes('Parent',h2);
title(axes2,'Angular velocity of payload vs time',
   'FontSize',16);
xlabel(axes2,'Time(s)');
ylabel(axes2,'Angular velocity (rad/s)');
hold(axes2,'all');
plot(tdata,w2data,'b');
grid on;
```

```
h2=figure(7);
set(h2,'Position',[800,200,500,300],...
      'Menu','none',...
      'NumberTitle','off',...
      'Name','5th layer sliding surface'...
);
axes2=axes('Parent',h2);
title(axes2,'5^{th} layer sliding surface vs time',
   'FontSize',16);
xlabel(axes2,'Time(s)');
ylabel(axes2,'s_5');
hold(axes2,'all');
plot(tdata,surf5data,'b');
grid on;

h2=figure(8);
set(h2,'Position',[800,200,500,300],...
      'Menu','none',...
      'NumberTitle','off',...
      'Name','4th layer sliding surface'...
);
axes2=axes('Parent',h2);
title(axes2,'4^{th} layer sliding surface vs time',
   'FontSize',16);
xlabel(axes2,'Time(s)');
ylabel(axes2,'s_4');
hold(axes2,'all');
plot(tdata,surf4data);
grid on;

h2=figure(9);
set(h2,'Position',[800,200,500,300],...
      'Menu','none',...
      'NumberTitle','off',...
```

```
       'Name','3rd layer sliding surface'...
);
axes2=axes('Parent',h2);
title(axes2,'3^{rd} layer sliding surface vs time',
   'FontSize',16);
xlabel(axes2,'Time(s)');
ylabel(axes2,'s_3');
hold(axes2,'all');
plot(tdata,surf3data);
grid on;

h2=figure(10);
set(h2,'Position',[800,200,500,300],...
       'Menu','none',...
       'NumberTitle','off',...
       'Name','2nd layer sliding surface'...
);
axes2=axes('Parent',h2);
title(axes2,'2^{nd} layer sliding surface vs time',
   'FontSize',16);
xlabel(axes2,'Time(s)');
ylabel(axes2,'s_2');
hold(axes2,'all');
plot(tdata,surf2data);
grid on;

h2=figure(11);
set(h2,'Position',[800,200,500,300],...
       'Menu','none',...
       'NumberTitle','off',...
       'Name','1st layer sliding surface'...
);
axes2=axes('Parent',h2);
title(axes2,'1^{st} layer sliding surface vs time',
```

```
   'FontSize',16);
xlabel(axes2,'Time(s)');
ylabel(axes2,'s_1');
hold(axes2,'all');
plot(tdata,surf1data);
grid on;

function f= controller(A,B,M,C,G,xend,s,v,th1,th2,w1,w2 )

P=[-2,-4,-8,-10,-16,-20];
z=place(A,B,P);
A=inv(M);
b1=A(1,1);
b2=A(2,1);
b3=A(3,1);
B=inv(M)*(-G-C*[v;w1;w2]);
f1=B(1,1);
f2=B(2,1);
f3=B(3,1);

s1=z(1)*(-xend+s)+z(2)*v;
s2=z(3)*th1*sign(th1*s1)+s1;
s3=z(4)*w1*sign(w1*s2)+s2;
s4=z(5)*th2*sign(th2*s3)+s3;
s5=z(6)*w2+s4;

ueq=-(z(1)*v+z(2)*f1+z(3)*w1+z(4)*f2+z(5)*w2+z(6)*f3)
    /(z(2)*b1+z(4)*b2+z(6)*b3);
usw=(-0.1*sign(s5)-s5)/(z(2)*b1+z(4)*b2+z(6)*b3);
f=ueq+usw;
```

Matlab codes to plot Figures 4.18-4.27

```
function HSMC
```

```
clear all;
close all;

len1=0.5;
len2=0.5;
mass=0.5;
mass1=0.25;
mass2=0.5;
g=9.81;

s=0;
v=0;
th1=0;
th2=0;
w1=0;
w2=0;
dw1=0;
dw2=0;
f=80;
t=0;
s_1=s;
v_1=v;
th_1=th1;
th_2=th2;
w_1=w1;
w_2=w2;
dv=0;
dv_1=dv;
dw_1=dw1;
dw_2=dw2;
f_1=f;
dt=0.01; %
xend=1; %
```

```
sdata=s;
vdata=v;
th1data=th1;
th2data=th2;
w1data=w1;
w2data=w2;
fdata=f;
tdata=t;
dw1data=dw1;
dw2data=dw2;
s1data=[0];
s2data=[0];
s3data=[0];
Sdata=[0];

for i=0:0.01:40
s=s_1;
v=v_1;
th1=th_1;
th2=th_2;
w1=w_1;
w2=w_2;
dv=dv_1;
dw1=dw_1;
dw2=dw_2;

[s1 s2 s3 S f_1] = controller( mass,mass1,mass2,len1,
len2,s_1,v_1,th_1,th_2,w_1,w_2,dw_1,dw_2,xend);

M=[mass+mass1+mass2 (mass1+mass2)*len1*cos(th1)
      mass2*len2*cos(th1);...
   (mass1+mass2)*len1*cos(th1) (mass1+mass2)*len1^2
      mass2*len1*len2*cos(th1-th2);...
```

```
    mass2*len2*cos(th2) mass2*len1*len2*cos(th1-th2)
      mass2*len2^2];
C=[0 -1*(mass1+mass2)*len1*w1*sin(th1) -1*mass2*
      len2*w2*sin(th2);...
  0 0 mass2*len1*len2*w1*sin(th1-th2);...
  0 -1*mass2*len1*len2*w1*sin(th1-th2) 0];
G=[0;(mass1+mass2)*g*len1*sin(th1);mass2*g*len2*sin(th2)];
T=[f_1;0;0];
ddq=inv(M)*(T-G-C*[v;w1;w2]);
dv_1=ddq(1);
dw_1=ddq(2);
dw_2=ddq(3);
y=[v;w1;w2]+dt*ddq;
v_1=y(1);
w_1=y(2);
w_2=y(3);
z=[s;th1;th2]+0.5*dt^2*ddq+dt*[v;w1;w2];
s_1=z(1);
th_1=z(2);
th_2=z(3);
t=t+dt;
tdata = [tdata,t];
fdata = [fdata,f_1];
vdata = [vdata,v_1];
th1data=[th1data,th_1];
th2data=[th2data,th_2];
w1data=[w1data,w_1];
w2data=[w2data,w_2];
sdata=[sdata,s_1];
s1data=[s1data s1];
s2data=[s2data s2];
s3data=[s3data s3];
Sdata=[Sdata S];
end
```

```
h1=figure(1);
set(h1,'Position',[100,200,500,300],...
      'Menu','none',...
      'NumberTitle','off',...
      'Name','Trolley position'...
);
axes1=axes('Parent',h1);
title(axes1,'Trolley position vs Time','FontSize',16);
xlabel(axes1,'Time(s)');
ylabel(axes1,'Trolley position(m)');
hold(axes1,'all');
plot(tdata,sdata);
grid on;

h2=figure(2);
set(h2,'Position',[800,200,500,300],...
      'Menu','none',...
      'NumberTitle','off',...
      'Name','Hook angle'...
);
axes2=axes('Parent',h2);
title(axes2,'Hook angle vs time','FontSize',16);
xlabel(axes2,'Time(s)');
ylabel(axes2,'Hook angle(rad)');
hold(axes2,'all');
plot(tdata,th1data,'r');
grid on;

h2=figure(3);
set(h2,'Position',[800,200,500,300],...
      'Menu','none',...
      'NumberTitle','off',...
      'Name','Payload angle'...
);
axes2=axes('Parent',h2);
```

```
title(axes2,'Payload angle vs time','FontSize',16);
xlabel(axes2,'Time(s)');
ylabel(axes2,'Payload angle(rad)');
hold(axes2,'all');
plot(tdata,th2data,'r');
grid on;

h2=figure(4);
set(h2,'Position',[800,200,500,300],...
      'Menu','none',...
      'NumberTitle','off',...
      'Name','Control input'...
);
axes2=axes('Parent',h2);
title(axes2,'Control input vs time','FontSize',16);
xlabel(axes2,'Time(s)');
ylabel(axes2,'Force(N)');
hold(axes2,'all');
plot(tdata,fdata,'r');
grid on;

h2=figure(5);
set(h2,'Position',[800,200,500,300],...
      'Menu','none',...
      'NumberTitle','off',...
      'Name','Angular velocity of hook'...
);
axes2=axes('Parent',h2);
title(axes2,'Angular velocity of hook vs time',
   'FontSize',16);
xlabel(axes2,'Time(s)');
ylabel(axes2,'Angular velocity(rad/s)');
hold(axes2,'all');
plot(tdata,w1data,'r');
grid on;
```

```
h2=figure(6);
set(h2,'Position',[800,200,500,300],...
      'Menu','none',...
      'NumberTitle','off',...
      'Name','Angular velocity of payload'...
);
axes2=axes('Parent',h2);
title(axes2,'Angular velocity of payload vs time',
   'FontSize',16);
xlabel(axes2,'Time(s)');
ylabel(axes2,'Angular velocity(rad/s)');
hold(axes2,'all');
plot(tdata,w2data,'r');
grid on;
h2=figure(7);
set(h2,'Position',[800,200,500,300],...
      'Menu','none',...
      'NumberTitle','off',...
      'Name','2nd layer sliding surface'...
);
axes2=axes('Parent',h2);
title(axes2,'The 2nd layer sliding surface S vs time',
   'FontSize',16);
xlabel(axes2,'Time(s)');
ylabel(axes2,'S');
hold(axes2,'all');
plot(tdata,Sdata,'r');
grid on;

h2=figure(8);
set(h2,'Position',[800,200,500,300],...
      'Menu','none',...
      'NumberTitle','off',...
      'Name','1st layer sliding surface s1'...
);
```

```
axes2=axes('Parent',h2);
title(axes2,'The 1st layer sliding surface s1 vs time',
   'FontSize',16);
xlabel(axes2,'Time(s)');
ylabel(axes2,'s1');
hold(axes2,'all');
plot(tdata,s1data,'r');
grid on;

h2=figure(9);
set(h2,'Position',[800,200,500,300],...
      'Menu','none',...
      'NumberTitle','off',...
      'Name','1st layer sliding surface s2'...
);
axes2=axes('Parent',h2);
title(axes2,'The 1st layer sliding surface s2 vs time',
   'FontSize',16);
xlabel(axes2,'Time(s)');
ylabel(axes2,'s2');
hold(axes2,'all');
plot(tdata,s2data,'r');
grid on;

h2=figure(10);
set(h2,'Position',[800,200,500,300],...
      'Menu','none',...
      'NumberTitle','off',...
      'Name','1st layer sliding surface s3'...
);
axes2=axes('Parent',h2);
title(axes2,'The 1st layer sliding surface s3 vs time',
   'FontSize',16);
xlabel(axes2,'Time(s)');
ylabel(axes2,'s3');
```

```
hold(axes2,'all');
plot(tdata,s3data,'r');
grid on;

function [s1 s2 s3 S f_1] = controller( mass,mass1,mass2,
     len1,len2,s_1,v_1,th_1,th_2,w_1,w_2,dw_1,dw_2,xend)
%UNTITLED Summary of this function goes here
% Detailed explanation goes here
A=(mass+mass1+mass2);
B=(mass1+mass2);
b=0;g=9.81;
s=s_1;
v=v_1;
th1=th_1;
th2=th_2;
w1=w_1;
w2=w_2;
dw1=dw_1;
dw2=dw_2;

M=[mass+mass1+mass2 (mass1+mass2)*len1*cos(th1)
     mass2*len2*cos(th1);...
  (mass1+mass2)*len1*cos(th1) (mass1+mass2)*len1^2
     mass2*len1*len2*cos(th1-th2);...
  mass2*len2*cos(th2) mass2*len1*len2*cos(th1-th2)
     mass2*len2^2];
C=[0 -1*(mass1+mass2)*len1*w1*sin(th1) -1*mass2*len2*
     w2*sin(th2);...
  0 0 mass2*len1*len2*w1*sin(th1-th2);...
  0 -1*mass2*len1*len2*w1*sin(th1-th2) 0];
G=[0;(mass1+mass2)*g*len1*sin(th1);mass2*g*len2*sin(th2)];
D=2*inv(M);
bi1=D(1,1);
bi2=D(2,1);
```

```
bi3=D(3,1);
B=inv(M)*(-G-C*[v;w1;w2]);
f1=B(1,1);
f2=B(2,1);
f3=B(3,1);

K=0.5;ante=0.02;
a1=10;a2=0.2;a3=-15;
b1=4.5;b2=0.05;b3=-0.25;
% first_level sliding surface
s1=v+a1*(s-xend);
s2=w1+a2*th1;
s3=w2+a3*th2;
% second_level sliding surface
S=b1*s1+b2*s2+b3*s3;

ueq=-(b1*a1*v+b1*f1+b2*a2*w1+b2*f2+b3*a3*w2+b3*f3)/
        (b1*bi1+b2*bi2+b3*bi3);
usw=-(K*S+ante*sign(S))/(b1*bi1+b2*bi2+b3*bi3);
f_1=(ueq+usw);
```

References

[1] Edwards C, Spurgeon S. Sliding mode control: Theory and applications. Padstow: CRC Press, 1998.

[2] Utkin V, Guldner J, Shi J. Sliding mode control in electro-mechanical systems. Padstow: CRC Press, 2009.

[3] Utkin VI. Sliding modes in control and optimization. Berlin: Springer, 2013.

[4] Dong Y, Wang Z, Feng Z, Cheng J. Incremental sliding mode control for double-pendulum-type overhead crane system. In: Proceedings of 27th Chinese Control Conference, Kunming, China, 2008, pp. 368–371.

[5] Wang W, Liu XD, Yi JQ. Structure design of two types of sliding-mode controllers for a class of under-actuated mechanical systems. IET Control Theory Appl, 2007, 1(1): 163–172.

[6] Tuan LA, Lee SG. Sliding mode controls of double-pendulum crane systems. J Mech Sci Technol, 2013, 27(6): 1863–1873.

[7] Qian D, Yi J. Hierarchical sliding mode control for under-actuated cranes. Berlin: Springer, 2016.

[8] Xue D, Chen Y, Atherton DP. Linear feedback control: Analysis and design with MATLAB. Philadelphia, PA: Society for Industrial and Applied Mathematics, 2007.

5 Output Feedback-Based Anti-sway Control

Abstract: This chapter focuses on the methodology of output feedback for the anti-sway control problem of double-pendulum-type overhead cranes. In practice, the angular velocities of hook and payload are unavailable in most cases. Such a practical issue calls for the output feedback-based control design. This chapter starts with some basic concepts about output feedback. Then, an output feedback-based controller is presented for double-pendulum-type overhead cranes. The stability analysis is also proven in the sense of Lyapunov. Simulation results are illustrated to support the control method. Finally, all the source codes are given.

Keywords: Output feedback-based design, system stability, double-pendulum-type motions, overhead cranes

5.1 Why is the Output Feedback Needed?

Control systems can be divided into two classes [1]. One class is based on state feedback. The other is based on output feedback. As suggested by its name, a state feedback-based control system employs all the system states to determine its control input. The method requires the knowledge of states. If any state is not available, a state observer has to be designed to estimate the system state for an observable system. The observer is employed to recover the state from measurements of inputs and outputs to the system. Consequently, such a control system integrates one controller and one observer. This kind of compound structure complicates the control system design. This increases the implementation cost and reduces the reliability of the control system. This is an inner drawback of the state feedback-based control design [2].

If the states are available for measurement, then the state feedback provides a simple way of designing a controller. Sometimes the state feedback becomes inevitable due to incomplete state information. In reality, most of the states are observable but they are immeasurable. So it is essential to find a controller based on the system output which is always measurable.

In order to overcome the drawback of the state feedback-based control design, the output feedback-based control design rises up. The output feedback control uses system outputs since outputs of a system are always measurable and available for feedback. The output feedback needs only the measurement of system output. The merit makes such a design available in many practical cases.

Recall the model of double-pendulum-type overhead crane systems (2.32). The model contains six state variables, such as the trolley position and its velocity, the hook angle and its angular velocity, the payload angle and its angular velocity. In many cases, the three velocity signals, such as the trolley velocity, the hook angular velocity, and the payload angular velocity, are unavailable [3].

DOI 10.1515/9783110520101-005

Though one can numerically differentiate position/angle signals to alternatively derive "velocities" in a numerical control system, a lot of noises will be induced due to many reasons, such as insufficient sensor accuracy, electromagnetic noises, and so on. If low-pass filters are adopted to further process the velocity signals, not only noises but also some useful signals will be filtered, and moreover, it is challenging to theoretically ensure that differentiating position/angle signals numerically along with low-pass filters can recover the exact velocities [4, 5].

As pointed out in Chapter 2, one property of the double-pendulum-type crane systems is underactuation. The sole control input is applied to the trolley such that both the hook and payload swings become underactuated. In order to stabilize the double-pendulum-type crane systems, the complicated coupling relationship between the actuated trolley translation and the underactuated swings has to be taken into consideration.

This chapter considers the output feedback-based double-pendulum-type crane control under the assumption that the trolley position, hook angle, and payload angle are measurable outputs. The assumption is challenging but practical. The challenge is that such a highly underactuated system has to be stabilized by the limited output variables. The reality is that only the limited output variables can be measurable in practice. Theoretically, how to analyze the stability of such an output feedback-based control system is also challenging because the double-pendulum-type overhead crane systems have the complicated nonlinearity and high underactuation degree.

5.2 Output Feedback Control Design and Stability Analysis

5.2.1 Control Design

Recall crane models (2.30) and (2.32), consider the crane control task and the assumption of the output feedback-based control design, and concern the crane operation. Then we can have the following conditions [4]:

$$|\theta_1| < \pi/2, \quad |\theta_2| < \pi/2, \quad l_1 \geq l_2 \tag{5.1}$$

$$\dot{x},\ \dot{\theta}_1, \text{and } \dot{\theta}_2 \text{ are unknown} \tag{5.2}$$

$$|u(t)| \leq u_{\max}. \tag{5.3}$$

$$\lim_{t\to\infty} x(t) = x_d,\ \lim_{t\to\infty} \theta_1(t) = 0,\ \text{and}\ \lim_{t\to\infty} \theta_2(t) = 0 \tag{5.4}$$

Equation (5.1) indicates that cables of the double-pendulum-type overhead crane systems can only endure the pulling tension. Equation (5.2) represents that the crane control design is based on the output feedback method. Equation (5.3) denotes that merely limited forces can be produced by the actuating motors owing to the physical constraints. The control task of the crane systems can be formulated by eq. (5.4).

Recall Lemma 3.2. The passivity of the double-pendulum-type crane systems can be adapted to realize the transport task. However, the passivity-based control law (3.29) contains the trolley velocity, which contradicts assumption (5.2).

Intuitionally, the control law (3.29) only contains the actuated trolley position and velocity while the important double-pendulum swing feedback-related signals, which are instrumental for fast swing damping, are not employed. To improve the control performance according to assumption (5.2), another Lyapunov has to be formulated in order to achieve the output feedback-based control design.

In order to simultaneously contain the actuated trolley and the underactuated hook and payload, a composite signal has to be constructed. To obtain the composite signal, we need to respect the dynamic characteristics of the crane model (2.30). For this purpose, integrate eq. (2.30) twice with respect to time and divide both sides with $m_0 + m_1 + m_2$. Then, eq. (5.5) can be obtained:

$$x + \frac{m_2 l_2}{m_t}[\beta \sin\theta_1 + \sin\theta_2] = \frac{1}{m_t}\iint u\,dt\,dt \tag{5.5}$$

Here $m_t = m_0 + m_1 + m_2$ and $\beta = (m_1 + m_2)l_1/m_2/l_2 > 1$.

The time derivative of eq. (5.5) has the form

$$\dot{x} + \frac{m_2 l_2}{m_t}[\beta\dot{\theta}_1\cos\theta_1 + \dot{\theta}_2\cos\theta_2] = \frac{1}{m_t}\int u dt \tag{5.6}$$

Define an energy-like function E_p

$$E_\mathrm{p} = \frac{1}{2m_t}\left(\int u dt\right)^2 \tag{5.7}$$

The time derivative of eq. (5.7), by inserting eq. (5.6), can be calculated as

$$\dot{E}_\mathrm{p} = \left[\dot{x} + \frac{m_2 l_2}{m_t}(\beta\dot{\theta}_1\cos\theta_1 + \dot{\theta}_2\cos\theta_2)\right]u \tag{5.8}$$

Replace $m_2 l_2/m_t$ with a tunable parameter α in eq. (5.8) to make the function E_p positive definite. Then, re-consider the left-hand side of eq. (5.5) and define the following composite signal [5]:

$$\phi = x + \alpha\,[\beta\sin\theta_1 + \sin\theta_2] \tag{5.9}$$

where $\alpha < 0$ is a predefined constant.

Consider the replacement. Accordingly, the energy-like function has to be modified as well, which is determined as

$$\dot{E}_\phi = \dot{x}u + \alpha u(\beta\dot{\theta}_1\cos\theta_1 + \dot{\theta}_2\cos\theta_2) \tag{5.10}$$

Define

$$\begin{aligned} \dot{E}_{\mathrm{m}} &= \dot{x}u \\ \dot{E}_{\mathrm{c}} &= \alpha u(\beta\dot{\theta}_1\cos\theta_1 + \dot{\theta}_2\cos\theta_2) \end{aligned} \tag{5.11}$$

Then, we have

$$\dot{E}_\phi = \dot{E}_{\mathrm{m}} + \dot{E}_{\mathrm{c}} \tag{5.12}$$

where E_{m} denotes the control input applied to the actuated trolley and E_{c} means the control input effecting the underactuated hook and payload by the inherent coupling.

The explicit expressions of E_{m} and E_{c} can also be drawn. Recall eq. (3.17). The physical background indicates that E_{m} is the total mechanical energy of the double-pendulum-type overhead crane systems. Therefore, the expression of E_{m} can be written as

$$E_{\mathrm{m}} = \frac{1}{2}\dot{\mathbf{q}}^T\mathbf{M}(\mathbf{q})\dot{\mathbf{q}} + (m_1 + m_2)gl_1(1 - \cos\theta_1) + m_2gl_2(1 - \cos\theta_2) \tag{5.13}$$

Further, the expression of E_{c} [5] has the form

$$\begin{aligned} E_{\mathrm{c}} = -\alpha \Bigg\{ & \frac{(m_1 + m_2)^2}{2m_2l_2} l_1^2\dot{\theta}_1^2\sin^2\theta_1 + \frac{1}{2}m_2l_2\dot{\theta}_2^2\sin^2\theta_2 + (m_1 + m_2)l_1\dot{\theta}_1\dot{\theta}_2\sin\theta_1\sin\theta_2 \\ & + \frac{m_0(m_1 + m_2)}{2m_2l_2}l_1^2\dot{\theta}_1^2 + \frac{m_0 + m_1}{2}l_2\dot{\theta}_2^2 + m_0l_1\dot{\theta}_1\dot{\theta}_2\cos(\theta_1 - \theta_2) \\ & + m_tg[\beta(1 - \cos\theta_1) + (1 - \cos\theta_2)]\Bigg\} \end{aligned} \tag{5.14}$$

Let

$$E_\phi = E_{\mathrm{m}} + E_{\mathrm{c}} \tag{5.15}$$

where E_{m} and E_{c} are, respectively given by eqs. (3.13) and (3.14).

Formulated by eq. (5.9), the composite signal is a nonlinear combination of x, θ_1, and θ_2. Such a composite signal can capture the dynamic characteristics of x, θ_1, and θ_2. Concerning the control task of the double-pendulum-type overhead crane systems, it is concluded that $\phi \rightarrow x_d$. Consequently, an error-like signal can be defined on the basis of the composite signal ϕ:

$$e_\phi = \phi - x_d = e_x + \alpha[\beta\sin\theta_1 + \sin\theta_2] \tag{5.16}$$

where $e_x = x - x_d$ is the positioning error. From eq. (5.16), it can be seen that the control task will be achieved by rendering $e_\phi \to 0$ with the forthcoming controller.

Concerning the error-like signal, some error-related terms are injected into E_ϕ in eq. (5.15) to obtain an energy-like function as

$$E_{e_\phi} = E_\phi + \frac{2k_p}{\pi}\left[e_\phi \arctan(e_\phi) - \frac{1}{2}\log(1 + e_\phi^2)\right] \tag{5.17}$$

Here k_p denotes a positive control parameter. Differentiating E_{e_ϕ} with respect to time gives

$$\dot{E}_{e_\phi} = \dot{e}_\phi\left[u + \frac{2k_p}{\pi}\arctan(e_\phi)\right] \tag{5.18}$$

If all the state variables are measurable, then the state feedback-based control law of the double-pendulum-type overhead crane systems can be designed as

$$u = -\frac{2k_p}{\pi}\arctan(e_\varphi) - \frac{2k_d}{\pi}\arctan(\dot{e}_\varphi) \tag{5.19}$$

Here k_d denotes another positive control parameter. Apparently, replace u in eq. (5.18) by eq. (5.19). Then, we can have

$$\dot{E}_{e_\phi} = -\frac{2k_p}{\pi}\dot{e}_\phi \arctan(\dot{e}_\phi) \le 0 \tag{5.20}$$

Unfortunately, the assumption in eq. (5.2) means the control law in eq. (5.19) is not available. To overcome this issue, the velocity feedback in eq. (5.19) should be replaced with some velocity-like signals. For this purpose, we construct another energy-like function based on eq. (5.17), formulated by

$$\begin{aligned} V = {} & E_\phi + \frac{2k_p}{\pi}\left[e_\phi \arctan(e_\phi) - \frac{1}{2}\log(1 + e_\phi^2)\right] \\ & + \frac{2}{\pi}\left\{(\chi + k_d e_\phi)\arctan(\chi + k_d e_\phi) - \frac{1}{2}\log[1 + (\chi + k_d e_\phi)^2]\right\} \end{aligned} \tag{5.21}$$

where χ is an auxiliary signal, generated by

$$\dot{\chi} + k_d \chi = -k_d^2 e_\phi \tag{5.22}$$

The derivative of V with respect to time can be obtained as

$$\begin{aligned} \dot{V} = {} & \dot{e}_\phi\left[u + \frac{2k_p}{\pi}\arctan(e_\phi) + \frac{2k_d}{\pi}\arctan(\chi + k_d e_\phi)\right] \\ & - \frac{2k_d}{\pi}(\chi + k_d e_\phi)\arctan(\chi + k_d e_\phi) \end{aligned} \tag{5.23}$$

According to eq. (5.23), design the output feedback-based control law as

$$u = -\frac{2k_p}{\pi}\arctan(e_\phi) - \frac{2k_d}{\pi}\arctan(\chi + k_d e_\phi) \tag{5.24}$$

where k_p and k_d are positive constants. Since $|\arctan(\bullet)| \le \pi/2$ always holds true, there exists

$$|u| \le k_p + k_d \tag{5.25}$$

Pick up $u_{\max} \ge k_p + k_d$. It implies that the control law (5.24) can respect assumption (5.3). From eqs. (5.16) and (5.24), the composite signal in eq. (5.16) adds swing-related feedback to the control law and hence the composite signal improves swing-related damping. Considering the proposed control law (5.24), it contains no velocity signals and no differentiation operations to position/swing-angle signals are required.

5.2.2 Stability Analysis

Theorem 5.1. E_ϕ in eq. (5.15) is positive definite.

Proof. Consider eq. (5.15). There are two terms, that is, E_{m} and E_{c}, involved in eq. (5.15). Proven in eq. (3.12), E_{m} indicates the total energy of the double-pendulum-type overhead crane systems. E_{m} is inherently positive definite. This case points out that E_ϕ is positive definite if E_{c} could be positive definite.

Considering the expression of E_{c} from eq. (5.14), the following equation is obtained:

$$\begin{aligned}\frac{E_{\mathrm{c}}}{\alpha} &= \frac{(m_1+m_2)^2}{2m_2l_2}l_1^2\dot{\theta}_1^2\sin^2\theta_1 - \frac{1}{2}m_2l_2\dot{\theta}_2^2\sin^2\theta_2 - (m_1+m_2)l_1\dot{\theta}_1\dot{\theta}_2\sin\theta_1\sin\theta_2\\ &\quad - \frac{m(m_1+m_2)}{2m_2l_2}l_1^2\dot{\theta}_1^2 - \frac{m+m_1}{2}l_2\dot{\theta}_2^2 - ml_1\dot{\theta}_1\dot{\theta}_2\cos(\theta_1-\theta_2)\\ &\quad - m_tg[\beta(1-\cos\theta_1) + (1-\cos\theta_2)]\\ &= E_{ck} - m_tg[\beta(1-\cos\theta_1) + (1-\cos\theta_2)]\end{aligned} \tag{5.26}$$

Further, E_{ck} can be deduced as

$$\begin{aligned}E_{ck} &\le -\left[\frac{(m_1+m_2)}{\sqrt{2m_2l_2}}l_1\dot{\theta}_1\sin\theta_1 + \sqrt{\frac{m_2l_2}{2}}\dot{\theta}_2\sin\theta_2\right]^2 - \frac{m_0l_1^2}{2l_2}\left|\dot{\theta}_1\right|^2 + m_1l_1\left|\dot{\theta}_1\right|\left|\dot{\theta}_2\right|\\ &= -\left[\frac{(m_1+m_2)}{\sqrt{2m_2l_2}}l_1\dot{\theta}_1\sin\theta_1 + \sqrt{\frac{m_2l_2}{2}}\dot{\theta}_2\sin\theta_2\right]^2 - \frac{m_0}{2}\left[\frac{l_1}{\sqrt{l_2}}\left|\dot{\theta}_1\right| - \sqrt{l_2}\left|\dot{\theta}_2\right|\right]^2\end{aligned} \tag{5.27}$$

Since $|\cos(\theta_1 - \theta_2)| \le 1$ in eq. (5.26), the following inequality can be deduced from eqs. (5.26) and (5.27):

$$E_c \ge -\alpha\left[\frac{(m_1+m_2)}{\sqrt{2m_2 l_2}} l_1\dot{\theta}_1 \sin\theta_1 + \sqrt{\frac{m_2 l_2}{2}}\dot{\theta}_2 \sin\theta_2\right]^2 - \frac{\alpha m_0}{2}\left[\frac{l_1}{\sqrt{l_2}}\left|\dot{\theta}_1\right| - \sqrt{l_2}\left|\dot{\theta}_2\right|\right]^2 - \alpha m_t g[\beta(1-\cos\theta_1) + (1-\cos\theta_2)] \tag{5.28}$$

As defined in eq. (5.9), the predefined constant α is negative. $E_c > 0$ can be deduced from eq. (5.28), that is, E_c is positive definite. Consequently, E_ϕ in eq. (5.15) is positive definite. Accordingly, other energy-like functions such as E_{e_ϕ} and V are also positive definite.

Theorem 5.2. The control law (5.24) always assures that $|u| \le u_{max}$, and the equilibrium point of the closed-loop system is of asymptotic stability, that is,

$$\lim_{t\to\infty}\left\|[e_x, \dot{x}, \theta_1, \dot{\theta}_1, \theta_2, \dot{\theta}_2]^T\right\| = 0 \tag{5.29}$$

Proof. Take the energy-like function V in eq. (5.21) into consideration as a Lyapunov function. According to Theorem 5.1, V is positive definite. Then, the derivative of V can be written as

$$\dot{V} = -\frac{2k_d}{\pi}\arctan(\chi + k_d e\phi_\phi)\arctan(\chi + k_d e_\phi) \le 0 \tag{5.30}$$

where $\dot{\Theta} \bullet \arctan(\Theta) \ge 0$ for any Θ.

From $e_x = x - x_d$, $\dot{e}_x = \dot{x}$. Further, considering eqs. (5.21) and (5.30), the following expressions can be drawn:

$$V \in L_\infty \Rightarrow e_\varphi, \chi, \dot{e}_x, \theta_1, \theta_2, e_x, x \in L_\infty \tag{5.31}$$

Considering $|u| \le u_{max}$ and substituting eq. (5.31) into the crane model (2.30) results in

$$\ddot{x}, \ddot{\theta}_1, \ddot{\theta}_2 \in L_\infty \tag{5.32}$$

Consequently, the closed-loop equilibrium point of the double-pendulum-type crane control system is stable in the sense of Lyapunov. In order to analyze the asymptotic stability of the closed-loop control system, a set is defined as

$$\Psi = \left\{(e_x, \theta_1, \theta_2, \dot{x}, \dot{\theta}_1, \dot{\theta}_2) : \dot{V} = 0\right\} \tag{5.33}$$

Let Ω be the largest invariant set in Ψ. Then, then, the following equations can be deduced from eq. (5.30):

$$\begin{aligned} \chi + k_d e_\phi &= 0 \\ \dot{\chi} + k_d \dot{e}_\phi &= 0 \end{aligned} \tag{5.34}$$

Equation (5.34) implies that eq. (5.35) holds true in the set Ψ, deduced from eqs. (5.23) and (5.24):

$$\begin{aligned} \dot{\chi} &= 0 \\ u &= -\frac{2k_p}{\pi}\arctan(e_\phi) \end{aligned} \tag{5.35}$$

Substituting eq. (5.35) into the second equation of eq. (5.34) gives

$$\dot{e}_\phi = 0 \Leftrightarrow \dot{e}_x + \alpha[\beta\dot{\theta}_1\cos\theta_1 + \dot{\theta}_2\cos\theta_2] = 0 \tag{5.36}$$

Thus, eq. (5.37) can be derived from eqs. (5.34) and (5.35), written as

$$e_\phi = \lambda_1 \Rightarrow u = -\frac{2k_p}{\pi}\arctan(\lambda_1) \tag{5.37}$$

with λ_1 being a constant to be determined. For clarity, the following proof is divided into two parts.

Part 1 presents that $e_x = 0$, $\dot{e}_x = 0$, $u(t) = 0$ in the largest invariant set Ω.

From eq. (5.36), we can have

$$\begin{aligned} \ddot{e}_\phi = 0 \Rightarrow{}& \alpha[(m_1+m_2)l_1(\ddot{\theta}_1\cos\theta_1 - \dot{\theta}_1^2\sin\theta_1) + m_2l_2\ddot{\theta}_2\cos\theta_2 - m_2l_2\dot{\theta}_2^2\sin\theta_2] \\ ={}& -m_2l_2\ddot{x} \end{aligned} \tag{5.38}$$

By comparing eq. (5.38) with eq. (2.30), we can conclude that

$$\alpha[u - m_t\ddot{x}] = -m_2l_2\ddot{x} \Rightarrow \ddot{x} = \frac{-2\alpha k_p\arctan(\lambda_1)}{\pi[\alpha m_t - m_2l_2]} \tag{5.39}$$

Consider eq. (5.37). Assuming $\lambda_1 \neq 0$, eq. (5.39) represents

$$\dot{e}_x = \dot{x} = \frac{-2\alpha k_p\arctan(\lambda_1)}{\pi[\alpha m_t - m_2l_2]}t + \lambda_2 \Rightarrow |\dot{e}_x| \to \infty \text{ as } t \to \infty, \lambda_2 \in R \tag{5.40}$$

with λ_2 being a constant. Obviously, the result in eq. (5.40) invalidates the following conclusion in eq. (5.31):

$$\dot{e}_x \in L_\infty$$

Consequently, $\lambda_1 = 0$. Thus, one can conclude from eqs. (5.37), (5.39), and (5.40) that

$$e_\phi = 0, u = 0, \ddot{x} = 0, \dot{e}_x = \lambda_2 \tag{5.41}$$

Similarly, assuming $\lambda_2 = 0$, we have $|e_x| = |\lambda_2 + \lambda_3| \to \infty$ when $t \to \infty$, with λ_3 being constant, reaching a contradiction with $e_x \in L_\infty$ in eq. (5.41). As a consequence, we have $\lambda_2 = 0$ and

$$\dot{e}_x = 0, \dot{x} = 0, e_x = \lambda_3 \tag{5.42}$$

Further, it is derived from eqs. (5.16), (5.41), and (5.42):

$$\begin{aligned} &\beta \sin\theta_1 + \sin\theta_2 = -\frac{\lambda_3}{\alpha} \\ &\beta\dot{\theta}_1 \cos\theta_1 + \dot{\theta}_2 \cos\theta_2 = 0 \end{aligned} \tag{5.43}$$

In the largest invariant set Ω, $\ddot{x} = 0$ from eq. (5.41) and substitute the condition in the crane model (2.30). Subsequently, recall $\beta = (m_1 + m_2)l_1/(m_2 l_2)$ in eq. (5.5). Equation (5.44) can be determined as

$$\begin{aligned} &\beta\ddot{\theta}_1 + \ddot{\theta}_2 \cos(\theta_1 - \theta_2) + \dot{\theta}_2^2 \sin(\theta_1 - \theta_2) = -\frac{g}{l_1}\beta \sin\theta_1 \\ &\ddot{\theta}_1 \cos(\theta_1 - \theta_2) + \frac{l_2}{l_1}\ddot{\theta}_2 - \dot{\theta}_1^2 \sin(\theta_1 - \theta_2) = -\frac{g}{l_1} \sin\theta_2 \end{aligned} \tag{5.44}$$

Add the two equations in eq. (5.44). Equation (5.45) can hold true:

$$\frac{\mathrm{d}}{\mathrm{d}t}\left[\beta\dot{\theta}_1 + \frac{l_2}{l_1}\dot{\theta}_2 + (\dot{\theta}_1 + \dot{\theta}_2)\cos(\theta_1 - \theta_2)\right] = \frac{\lambda_3 g}{\alpha l_1} \tag{5.45}$$

Integrating both sides of eq. (5.45) yields

$$\beta\dot{\theta}_1 + \frac{l_2}{l_1}\dot{\theta}_2 + (\dot{\theta}_1 + \dot{\theta}_2)\cos(\theta_1 - \theta_2) = \frac{\lambda_3 g}{\alpha l_1} t + \lambda_4 \tag{5.46}$$

where λ_4 is a constant to be determined.

Similarly, if it is assumed that $\lambda_3 \neq 0$, the left-hand side of eq. (5.46) will be infinite as $t \to \infty$. Since the derivatives of θ_1 and θ_2 are bounded according to eq. (5.41), the left-hand side in eq. (5.46) must be bounded as well. The assumption leads to an apparent contradiction. Hence, eq. (5.47) can be deduced from eqs. (5.41) and (5.42), in the form

$$\lambda_3 = 0 \Rightarrow e_x = 0, \beta \sin\theta_1 = -\sin\theta_2 \text{ and } \beta\dot{\theta}_1 \cos\theta_1 = -\dot{\theta}_2 \cos\theta_2 \tag{5.47}$$

Part 2 presents that θ_1, θ_2 and their derivatives are equal to zero in the largest invariant set Ω.

Expanding $\cos(\theta_1 - \theta_2)$ gives

$$(\dot{\theta}_1 + \dot{\theta}_2)\cos(\theta_1 - \theta_2) = \dot{\theta}_1 \cos\theta_1 \cos\theta_2 + \dot{\theta}_2 \sin\theta_1 \sin\theta_2 + \dot{\theta}_1 \sin\theta_1 \sin\theta_2 + \dot{\theta}_2 \cos\theta_1 \cos\theta_2 \tag{5.48}$$

Considering eq. (5.47), eq. (5.48) can be rearranged as

$$\begin{aligned}(\dot{\theta}_1 + \dot{\theta}_2)\cos(\theta_1 - \theta_2) &= -\frac{1}{\beta}\dot{\theta}_2(\cos^2\theta_2 + \sin^2\theta_2) - \beta\dot{\theta}_1(\cos^2\theta_1 + \sin^2\theta_1)\\ &= -\frac{1}{\beta}\dot{\theta}_2 - \beta\dot{\theta}_1\end{aligned} \tag{5.49}$$

Note that $\sin^2\theta_i + \cos^2\theta_i \equiv 1$ $(i = 1, 2)$ in eq. (5.49).

Substituting $\lambda_3 = 0$ in eqs. (5.47) and (5.49) into eq. (5.46) yields

$$\frac{m_1}{\beta m_2}\dot{\theta}_2 = \lambda_4 \Rightarrow \ddot{\theta}_2 = 0 \tag{5.50}$$

The analysis indicates that $\dot{\theta}_1 \equiv 0 \Longleftrightarrow \dot{\theta}_2 \equiv 0$ in the largest invariant set Ω. Further, we will analyze that only one of the following two equations can hold true:

$$\dot{\theta}_1(t) \equiv 0 \text{ and } \dot{\theta}_2(t) \equiv 0 \tag{5.51}$$

$$\dot{\theta}_1(t) \not\equiv 0 \text{ and } \dot{\theta}_2(t) \not\equiv 0 \tag{5.52}$$

Assuming that eq. (5.52) holds true, $\lambda_4 \not\equiv 0$ can be deduced from eqs. (5.50) and (5.52). Substituting $\ddot{\theta}_2 = 0$ of eq. (5.50) into eqs. (5.43) and (5.44) yields

$$\beta\ddot{\theta}_1 + \dot{\theta}_2^2\sin(\theta_1 - \theta_2) = -\frac{g}{l_1}\beta\sin\theta_1 \tag{5.53}$$

$$\ddot{\theta}_1\cos(\theta_1 - \theta_2) - \dot{\theta}_1^2\sin(\theta_1 - \theta_2) = -\frac{g}{l_1}\sin\theta_2 \tag{5.54}$$

From eqs. (5.53) and (5.54), eq. (5.55) can be obtained:

$$\sin(\theta_1 - \theta_2)[\dot{\theta}_2^2\cos(\theta_1 - \theta_2) + \beta\dot{\theta}_1^2] \equiv \frac{g\beta}{l_1}[\sin\theta_2 - \cos(\theta_1 - \theta_2)\sin\theta_1] \tag{5.55}$$

Further, the right-hand side of eq. (5.55) can be rearranged as

$$\begin{aligned}\sin\theta_2 - \cos(\theta_1 - \theta_2)\sin\theta_1 &= \sin\theta_2 - \sin\theta_1\cos\theta_1\cos\theta_2 - \sin^2\theta_1\sin\theta_2\\ &= -\cos\theta_1[\sin\theta_1\cos\theta_2 - \cos\theta_1\sin\theta_2] = -\cos\theta_1\sin(\theta_1 - \theta_2)\end{aligned} \tag{5.56}$$

Consequently, eq. (5.55) can be rewritten as

$$\sin(\theta_1 - \theta_2)[\dot{\theta}_2^2\cos(\theta_1 - \theta_2) + \beta\dot{\theta}_1^2] \equiv -\frac{g\beta}{l_1}\cos\theta_1\sin(\theta_1 - \theta_2) \tag{5.57}$$

In order to obtain the expected conclusion, we have to illustrate that $\sin(\theta_1 - \theta_2) \not\equiv 0$. For this purpose, it is assumed that $\sin(\theta_1 - \theta_2) \equiv 0$ and eq. (5.58) can be deduced from the assumption

$$\sin(\theta_1 - \theta_2) = \sin\theta_1(\cos\theta_2 + \beta\cos\theta_1) \equiv 0 \tag{5.58}$$

Due to $\dot{\theta}_1(t) \not\equiv 0$ in eq. (5.52), it can be seen that $\sin\theta_1 \not\equiv 0$. Hence, eq. (5.59) can be drawn from eq. (5.58):

$$\cos\theta_2 \equiv -\beta\cos\theta_1 \Rightarrow \cos^2\theta_2 \equiv \beta^2\cos^2\theta_1 \tag{5.59}$$

From eq. (5.47), $\sin^2\theta_2 = \beta^2\sin^2\theta_1$ can be drawn. Adding it to eq. (5.59) gives

$$\beta^2\cos^2\theta_1\sin^2\theta_2 + \cos^2\theta_2 \equiv \beta^2(\sin^2\theta_1 + \cos^2\theta_1) \Rightarrow \beta^2 \equiv 1 \tag{5.60}$$

The result in eq. (5.60) contradicts the condition $\beta > 1$ in eq. (5.5).

Consequently, the assumption $\sin(\theta_1 - \theta_2) \equiv 0$ is false. Accordingly, the opposite side of the assumption is true, that is,

$$\sin(\theta_1 - \theta_2) \not\equiv 0 \tag{5.61}$$

Considering eq. (5.61), eq. (5.62) can be deduced from eq. (5.57) as follows:

$$\dot{\theta}_2^2\cos(\theta_1 - \theta_2) + \beta\dot{\theta}_1^2 \equiv -\frac{g\beta}{l_1}\cos\theta_1 \tag{5.62}$$

Further, substituting eq. (5.50) into eq. (5.47) shows

$$\dot{\theta}_1\cos\theta_1 = -\frac{m_2}{m_1}\lambda_4\cos\theta_2 \Rightarrow \dot{\theta}_1^2\cos^2\theta_1 = \left(\frac{m_2}{m_1}\right)^2\lambda_4^2\cos^2\theta_2 \tag{5.63}$$

Inserting eq. (5.63) into eq. (5.62) and multiplying with $m_1^2\cos^2\theta_1\cos^2\theta_2/m_2^2$ yields

$$\left(\frac{m_1}{m_2}\right)^2\dot{\theta}_2^2\cos^2\theta_1\cos^2\theta_2\cos(\theta_1 - \theta_2) + \frac{m_1^2\beta g}{m_2^2 l_1}\cos^3\theta_1\cos^2\theta_2 + \beta\lambda_4^2\cos^4\theta_2 \equiv 0 \tag{5.64}$$

Noting $\dot{\theta}_2^2\cos^2\theta_2 = \beta^2\dot{\theta}_1^2\cos^2\theta_1$ in eq. (5.47) and considering eq. (5.63), eq. (5.64) can be reduced as

$$\beta\lambda_4^2\cos^2\theta_1\cos^2\theta_2\cos(\theta_1 - \theta_2) + \frac{m_1^2 g}{m_2^2 l_1}\cos^3\theta_1\cos^2\theta_2 + \lambda_4^2\cos^4\theta_2 \equiv 0 \tag{5.65}$$

From eq. (5.47), we deal with $\cos^2\theta_2$ and $\cos(\theta_1-\theta_2)$ and obtain the following formulas:

$$\begin{aligned}\cos^2\theta_2 &= 1-\beta^2+\beta^2\cos^2\theta_1\\ \cos(\theta_1-\theta_2) &= \cos\theta_1\cos\theta_2-\beta+\beta\cos^2\theta_1\end{aligned}\tag{5.66}$$

Substituting eq. (5.66) into eq. (5.65) and rearranging eq. (5.65) yields

$$\begin{aligned}&\beta\lambda_4^2\cos^3\theta_1(1-\beta^2+\beta^2\cos^2\theta_1)\cos\theta_2\\ &\equiv -\lambda_4^2\beta^2(\cos^2\theta_1-1)\cos^2\theta_1(1-\beta^2+\beta^2\cos^2\theta_1)-\lambda_4^2(1-\beta^2+\beta^2\cos^2\theta_1)^2\\ &\quad-\frac{m_1^2g}{m_2^2l_1}\cos^3\theta_1(1-\beta^2+\beta^2\cos^2\theta_1)\end{aligned}\tag{5.67}$$

Note that the left-hand side of eq. (5.67) still contains a $\cos\theta_2$-related term. Fortunately, the left-hand side of eq. (5.66) is $\cos^2\theta_2$ and the right-hand side of eq. (5.66) is $\cos\theta_1$ related. This fact indicates that $\cos\theta_2$ in eq. (5.67) can be eliminated by squaring both sides of eq. (5.67). Then, the following expression purely with respect to $\cos\theta_1$ can be drawn:

$$\begin{aligned}&\beta^2\lambda_4^4\cos^6\theta_1(1-\beta^2+\beta^2\cos^2\theta_1)^3-(1-\beta^2+\beta^2\cos^2\theta_1)^2[\lambda_4^2\beta^2(\cos^2\theta_1-1)\cos^2\theta_1\\ &+\lambda_4^2(1-\beta^2+\beta^2\cos^2\theta_1)+\frac{m_1^2g}{m_2^2l_1}\cos^3\theta_1\Bigg]^2\equiv 0\end{aligned}\tag{5.68}$$

Apparently, eq. (5.68) is a 12th-order polynomial one-variable equation with respect to $\cos\theta_1$. It can be rewritten as

$$\sum_{i=0}^{12}\delta_i\cos^i\theta_1\equiv 0\tag{5.69}$$

Here δ_i is the corresponding real coefficient. Since an nth-order polynomial equation has no more than n real roots, $\cos\theta_1$ has at most 12 discrete values (real roots) under the assumption that a part of or all δ_i in eq. (5.69) are nonzero. On the other hand, because $\dot{\theta}_1\not\equiv 0$ is assumed in eq. (5.52), $\cos\theta_1$ is not constant. Meanwhile, $\theta_1\in L_\infty$ in eq. (5.31) indicates that θ_1 and $\cos\theta_1$ are uniformly continuous, that is, $\cos\theta_1$ varies continuously with time and has countless different values.

The result that $\cos\theta_1$ is not constant from the assumption that a part of or all δ_i in eq. (5.69) are nonzero contradicts the fact that $\cos\theta_1$ varies continuously with time and has countless different values. Consequently, the unique case deduced from such a contradiction is that all the coefficients δ_i in eq. (5.69) are zero in the sense that

$$\delta_0=\delta_1=\cdots=\delta_{12}\equiv 0\tag{5.70}$$

Take the coefficient of $\cos^0\theta_1$ in eq. (5.68), that is, δ_0, into consideration. δ_0 have to satisfy eq. (5.71) due to eq. (5.70)

$$\delta_0 = -\lambda_4^4(1-\beta^2)^4 \equiv 0 \tag{5.71}$$

Since $\lambda_4 \not\equiv 0$ can be deduced from assumption (5.52), the only solution of eq. (5.71) is

$$\beta^2 \equiv 1 \tag{5.72}$$

Unfortunately, the result contradicts the fact that $\beta > 1$ in eq. (5.5). Recall that the conclusion in eq. (5.72) is due to $\lambda_4 \not\equiv 0$, which stems from the assumed precondition $\dot{\theta}_1 \not\equiv 0$ and $\dot{\theta}_2 \not\equiv 0$ in eq. (5.52). The contradiction reveals that cases $\dot{\theta}_1 \not\equiv 0$ and $\dot{\theta}_2 \not\equiv 0$ do not exist. As a result, it is evident that, only the precondition $\dot{\theta}_1(t) \equiv 0$ and $\dot{\theta}_2(t) \equiv 0$ in eq. (5.51) holds true in the largest invariant set Ω in the sense that

$$\dot{\theta}_1(t) \equiv 0, \quad \dot{\theta}_2(t) \equiv 0 \quad \text{and } \lambda_4 = 0 \Rightarrow \ddot{\theta}_1 = 0 \tag{5.73}$$

Further, substituting eqs. (5.50) and (5.73) into eq. (5.44) yields

$$\begin{aligned} &-\beta \sin\theta_1 = 0 \Rightarrow \theta_1 = 0 \\ &-\frac{g}{l_1}\sin\theta_2 = 0 \Rightarrow \theta_2 = 0 \end{aligned} \tag{5.74}$$

In one sentence, eqs. (5.41), (5.42), (5.47), (5.50), (5.73), and (5.74) indicate that closed-loop equilibrium is what is merely contained in the largest invariant set Ω in terms of LaSalle's invariance principle [6], that is,

$$\Omega = \left\{(e_x, \theta_1, \theta_2, \dot{x}, \dot{\theta}_1, \dot{\theta}_2) : e_x, \theta_1, \theta_2, \dot{x}, \dot{\theta}_1, \dot{\theta}_2 = 0\right\} \tag{5.75}$$

5.3 Simulation Results

In order to demonstrate the effectiveness of the output feedback control design for double-pendulum-type overhead crane systems, some numerical simulation results are included by such a crane system. Some physical parameters of the crane [4] are configured as $m_0 = 10$ kg, $m_1 = 25$ kg, $m_2 = 40$ kg, $l_1 = 3$ m, $l_2 = 2$ m, and $g = 9.8$ m/s^2.

To achieve the transport task, the trolley destination from its initial position, that is, the origin, is set to be $x_d = 1$ m. In addition, we assume that the largest allowable permitted force constraint for the control input is assumed to be $u_{max} = 160$ N. The gains for the designed output feedback control law are conveniently selected as $k_p = 40$, $k_d = 120$, $\alpha = -0.06$.

As displayed in Figure 5.1, the designed output feedback controller is able to achieve the transport control objective. The trolley is stabilized at the desired position, that is, $x_d = 1$ m. The angles of payload and hook are illustrated in Figures 5.3 and 5.5. The curves indicate that the payload and hook are stabilized at the desired position. The control force applied to the trolley is displayed in Figure 5.7. The angular

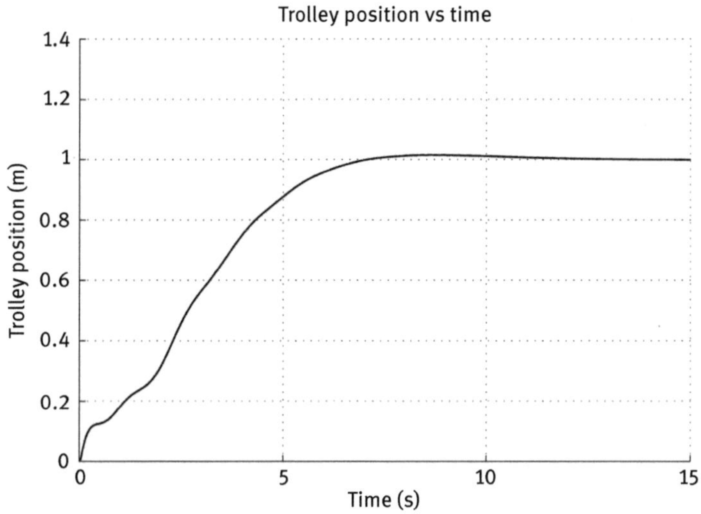

Figure 5.1: Trolley position versus time by the output feedback-based anti-sway control.

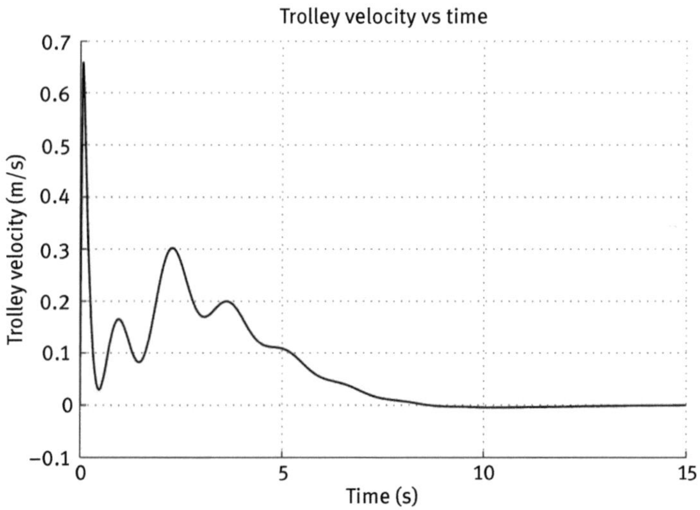

Figure 5.2: Trolley velocity versus time by the output feedback-based anti-sway control.

velocities of payload and hook are shown in Figures 5.4 and 5.6. From Figures 5.1 to 5.7, we find that the proposed control design can drive the trolley to the desired position accurately. At the same time, such a control design can eliminate the double-pendulum swing angles in an efficient manner.

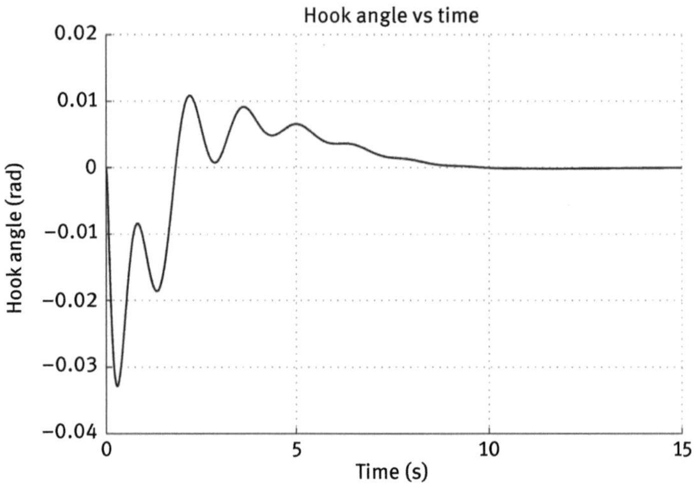

Figure 5.3: Hook angle versus time by the output feedback-based anti-sway control.

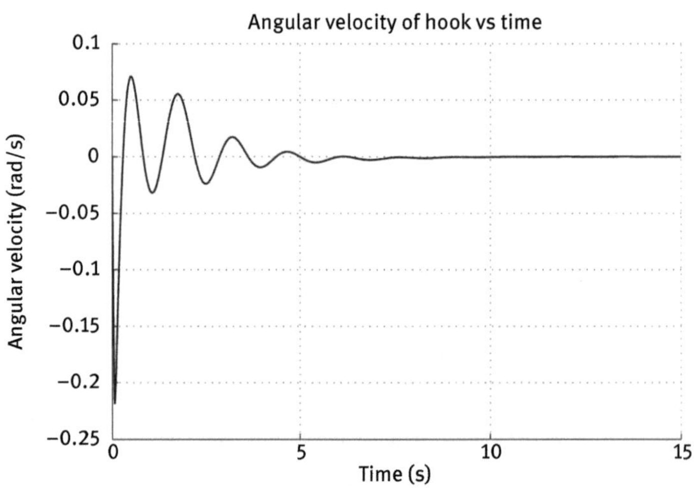

Figure 5.4: Angular velocity of hook versus time by the output feedback-based anti-sway control.

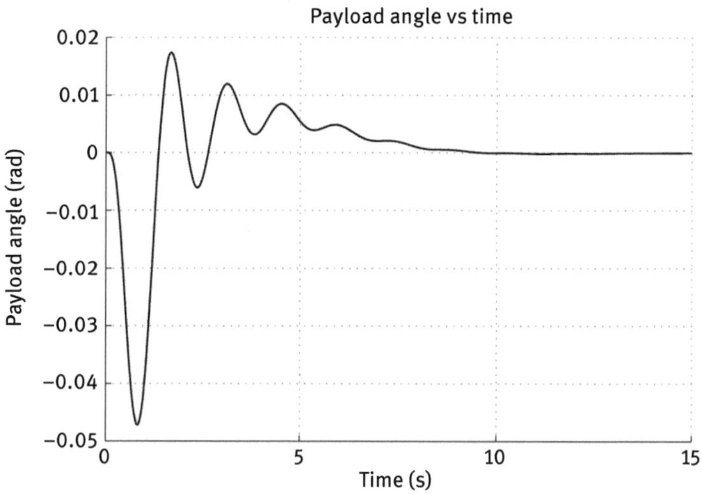

Figure 5.5: Payload angle versus time by the output feedback-based anti-sway control.

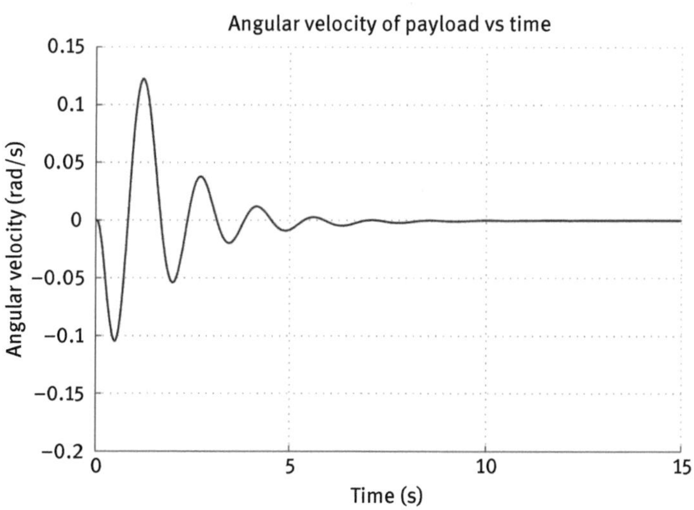

Figure 5.6: Angular velocity of payload versus time by the output feedback-based anti-sway control.

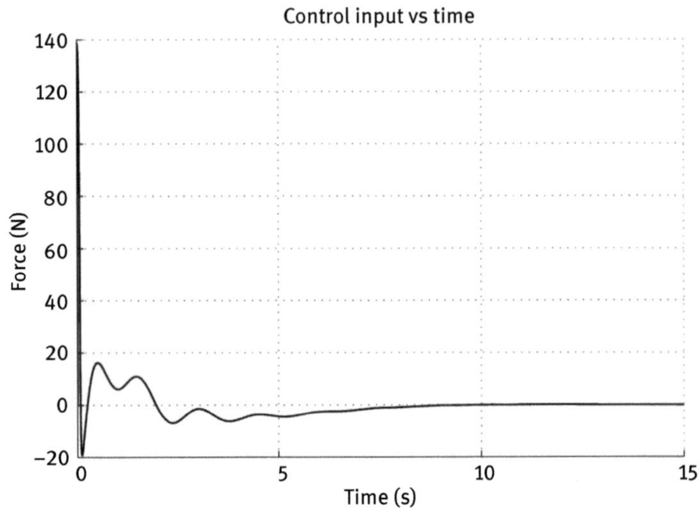

Figure 5.7: Control input versus time by the output feedback-based anti-sway control.

5.4 Source Codes

Figures 5.1–5.7 can be numerically obtained using Matlab software [7]. The source codes are listed as follows:

Matlab codes to plot Figures 5.1–5.7

```
function Outputfeedback
clear all;
close all;

%physical parameters
len1=3;
len2=2;
mass=10;
mass1=25;
mass2=40;
g=9.8;

%initial condition
s=0;
v=0;
```

```
th1=0;
th2=0;
w1=0;
w2=0;
f=0;
t=0;
dt=0.01;
xend=1; % desired position

sdata=s;
vdata=v;
th1data=th1;
th2data=th2;
w1data=w1;
w2data=w2;
fdata=f;
tdata=t;

s_1=s;
v_1=v;
th_1=th1;
th_2=th2;
w_1=w1;
w_2=w2;
x_1=0;
j=1;

for i=1:1:1500
s=s_1;
v=v_1;
th1=th_1;
th2=th_2;
w1=w_1;
w2=w_2;
```

```
x0=x_1;

c1=cos(th1);
c2=cos(th2);
s1=sin(th1);
s2=sin(th2);
c12=cos(th1-th2);
s12=sin(th1-th2);
j=j+1;

% crane model
M=[mass+mass1+mass2 (mass1+mass2)*len1*cos(th1) mass2*
       len2*cos(th2);...
  (mass1+mass2)*len1*cos(th1) (mass1+mass2)*len1^2
  mass2*len1*len2*cos(th1-th2);...
  mass2*len2*cos(th2) mass2*len1*len2*cos(th1-th2)
       mass2*len2^2];

C=[0 -1*(mass1+mass2)*len1*w1*sin(th1) -1*mass2*len2*
       w2*sin(th2);...
  0 0 mass2*len1*len2*w1*sin(th1-th2);...
  0 -1*mass2*len1*len2*w1*sin(th1-th2) 0];
G=[0;(mass1+mass2)*g*len1*sin(th1);mass2*g*len2*sin(th2)];

[F x_1]=controller(s,th1,th2,mass1,mass2,len1,len2,
       j,x0,xend);

f=xianzhi(F);

% apply controller to crane
T=[f;0;0];
q=inv(M)*(T-G-C*[v;w1;w2]);
y=[v;w1;w2]+dt*q;
v_1=y(1);
```

```
w_1=y(2);
w_2=y(3);
z=[s;th1;th2]+0.5*dt^2*q+dt*[v;w1;w2];
s_1=z(1);
th_1=z(2);
th_2=z(3);
t=t+dt;

tdata=[tdata,t];
fdata=[fdata,f];
vdata=[vdata,v_1];
th1data=[th1data,th_1];
th2data=[th2data,th_2];
w1data=[w1data,w_1];
w2data=[w2data,w_2];
sdata=[sdata,s_1];
end

h1=figure(1);
set(h1,'Position',[100,200,500,300],...
       'Menu','none',...
       'NumberTitle','off',...
       'Name','Trolley position'...
);
axes1=axes('Parent',h1);
title(axes1,'Trolley position vs time','FontSize',16);
xlabel(axes1,'Time(s)');
ylabel(axes1,'Trolley position(m)');
hold(axes1,'all');
plot(tdata,sdata);
grid on;

h1=figure(2);
set(h1,'Position',[100,200,500,300],...
```

```
        'Menu','none',...
        'NumberTitle','off',...
        'Name','Trolley velocity'...
);
axes1=axes('Parent',h1);
title(axes1,'Trolley velocity vs time','FontSize',16);
xlabel(axes1,'Time(s)');
ylabel(axes1,'Trolley velocity(m/s)');
hold(axes1,'all');
plot(tdata,vdata);
grid on;

h2=figure(3);
set(h2,'Position',[800,200,500,300],...
        'Menu','none',...
        'NumberTitle','off',...
        'Name','Hook angle'...
);
axes2=axes('Parent',h2);
title(axes2,'Hook angle vs time','FontSize',16);
xlabel(axes2,'Time(s)');
ylabel(axes2,'Hook angle(rad)');
hold(axes2,'all');
plot(tdata,th1data,'r');
grid on;

h2=figure(4);
set(h2,'Position',[800,200,500,300],...
        'Menu','none',...
        'NumberTitle','off',...
        'Name','Angular velocity'...
);
axes2=axes('Parent',h2);
title(axes2,'Angular velocity of hook vs time',
    'FontSize',16);
```

```
xlabel(axes2,'Time(s)');
ylabel(axes2,'Angular velocity(rad/s)');
hold(axes2,'all');
plot(tdata,w1data,'r');
grid on;

h2=figure(5);
set(h2,'Position',[800,200,500,300],...
       'Menu','none',...
       'NumberTitle','off',...
       'Name','Payload angle'...
);
axes2=axes('Parent',h2);
title(axes2,'Payload angle vs time','FontSize',16);
xlabel(axes2,'Time(s)');
ylabel(axes2,'Payload angle(rad)');
hold(axes2,'all');
lot(tdata,th2data,'r');
grid on;

h2=figure(6);
set(h2,'Position',[800,200,500,300],...
       'Menu','none',...
       'NumberTitle','off',...
       'Name','Angular velocity'...
);
axes2=axes('Parent',h2);
title(axes2,'Angular velocity of payload vs time',
   'FontSize',16);
xlabel(axes2,'Time(s)');
ylabel(axes2,'Angular velocity(rad/s)');
hold(axes2,'all');
plot(tdata,w2data,'r');
grid on;
```

```
h2=figure(7);
set(h2,'Position',[800,200,500,300],...
       'Menu','none',...
       'NumberTitle','off',...
       'Name','Control input'...
);
axes2=axes('Parent',h2);
title(axes2,'Control input vs time','FontSize',16);
xlabel(axes2,'Time(s)');
ylabel(axes2,'Force(N)');
hold(axes2,'all');
plot(tdata,fdata,'r');
grid on;

% output feedback-based controller
function [F x_1]= controller(s,th1,th2,mass1,mass2,len1,
   len2,j,x0,xend)
%
a=-0.01;
b=(mass1+mass2)*len1/(mass2*len2);
e=s-xend+a*(b*sin(th1)+sin(th2));

% composite signal
kd=120;
tspan=[0:0.001:0.01];
[ts,ix]=ode45(@(ts,x)dxdt(ts,x,e),tspan,x0);
ix;
x=ix(8,:);
x_1=x;

kp=40;
F=-2*kp*atan(e)/pi-2*kd*atan(x+kd*e)/pi;

% limited control input
function f= xianzhi(F)
```

```
if (F>160)
    f=160;
else if(F>0)
            f=F;
 else if (F<-160)
            f=-160;
      else
            f=F;
      end
   end
end

function dx=dxdt(ts,x,e)
dx=-120*x-120^2*e;
```

References

[1] Skogestad S, Postlethwaite I. Multivariable feedback control: Analysis and design. New York: Wiley, 2007.

[2] Zhou K, Doyle JC, Glover K. Robust and optimal control. Upper Saddle River, NJ: Prentice Hall, 1996.

[3] Kim D, Singhose W. Performance studies of human operators driving double-pendulum bridge cranes. Control Eng Practice, 2010, 18(6): 567–576.

[4] Sun N, Fang YC, Chen H, Lu B. Energy-based control of double pendulum cranes. In: Proceedings of IEEE Annual International Conference on Cyber Technology in Automation Control and Intelligent Systems, Shenyang, China, 2015, pp. 258–263.

[5] Sun N, Fang YC, Chen H, Lu B. Amplitude-saturated nonlinear output feedback antiswing control for underactuated cranes with double-pendulum cargo dynamics. IEEE Trans Ind Electron, 2017, 64(3): 2135–2146.

[6] Khalil HK. Nonlinear systems. Upper Saddle River, NJ: Prentice-Hall, 1996.

[7] Xue D, Chen Y, Atherton DP. Linear feedback control: Analysis and design with MATLAB. Philadelphia, PA: Society for Industrial and Applied Mathematics, 2007.

6 Fuzzy-Logic-Based Anti-sway Control

Abstract: This chapter focuses on the methodology of fuzzy logic for the anti-sway control problem of double-pendulum-type overhead cranes. Concerning control applications, fuzzy-logic-based control (short for fuzzy control) has been proven to be effective for solving unmodeled dynamics. However, fuzzy control needs a fuzzy inference model to determine the final control signal. In real applications, fuzzy control often suffers from the dilemma of exponential rule explosion. Fortunately, a fuzzy inference model called single-input-rule modules (SIRMs) can overcome the dilemma by an ingenious structure. This chapter starts with some basic concepts about fuzzy logic and fuzzy control. Then, SIRMs-based fuzzy controller is presented for double-pendulum-type overhead cranes. Some simulation results are illustrated to support the control method. Finally, all the source codes are listed.

Keywords: Fuzzy logic-based control design, single-input-rule modules, double-pendulum-type motions, overhead cranes

6.1 Fuzzy Logic and Fuzzy Control

Fuzzy logic is a logic that refers to the study of methods and principles of human reasoning [1]. The classical logic is two valued, that is, "true" or "false." However, many concepts are hardly described by such a simple logic. The term "fuzzy" refers to the fact that the logic involved can deal with concepts that cannot be expressed as "partially true" or "partially false."

Fuzzy logic was first proposed by Lotfi A. Zadeh of the University of California at Berkeley in 1965 [2]. Its ultimate goal is to provide foundations for approximate reasoning using imprecise propositions based on fuzzy set theory. Concerning the control applications of fuzzy logic, fuzzy logic has been widely used in the control areas because conventional control systems theory does not attempt to study any formulation, analysis, and control of what has been called fuzzy systems, which may be vague, incomplete, linguistically described, or even inconsistent.

Nowadays, fuzzy logic has the advantage that the solution to a problem can be cast in terms that human operators can understand. This makes it easier to mechanize tasks that are already successfully performed by humans. The implementation of fuzzy logic in the control areas raises the concept of fuzzy control, where fuzzy set theory plays a central role in the investigation of control design and implementation.

The majority of fuzzy control systems are knowledge-based systems in that either their fuzzy models or their fuzzy logic controllers are described by fuzzy inference systems. A fuzzy inference system contains fuzzifier, fuzzy inference engine, knowledge base, and defuzzifier, as illustrated in Figure 6.1.

DOI 10.1515/9783110520101-006

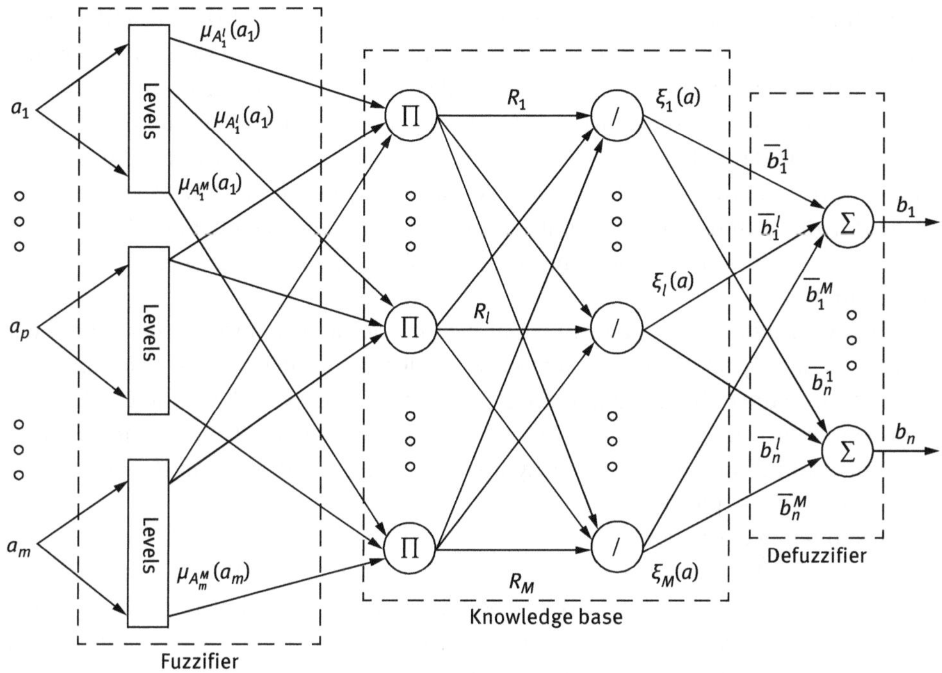

Figure 6.1: Prototype of the fuzzy inference system.

From Figure 6.1, the input vector of the fuzzy inference system is fed by $\mathbf{a} = [a_1\ a_2 \ldots a_m]^T \in \boldsymbol{R}^m$, the output vector is marked by $\mathbf{b} = [b_1\ b_2 \ldots b_n]^T \in \boldsymbol{R}^n$, and its IF–THEN rules are expressed by $R = \cup_{l=1}^{M} R_l$, where R means a set of fuzzy rules and M is the number of rules. R_l in the set R is depicted by

$$R_l : \text{ IF } a_1 \text{ is } A_1^l \text{ and } a_2 \text{ is } A_2^l \ldots . a_m \text{ is } A_m^l, \text{ THEN } b_1 \text{ is } B_1^l \ldots . \text{ and } b_n \text{ is } B_n^l \tag{6.1}$$

where A_p^l and B_q^l are the linguistic variables ($p = 1, 2, \ldots, m$ and $q = 1, 2, \ldots, n$). Both A_p^l and B_q^l are formulated by their membership functions $\mu_{A_p^l}(a_p)$ and $\mu_{B_q^l}(b_q)$ ($l = 1, 2, \ldots, M$).

The fuzzy inference system adapts the singleton fuzzifier, the center average defuzzifier, and the product inference. Finally, the qth output is calculated as

$$b_q = \sum_{l=1}^{M} \bar{b}_q^l \xi_l(\mathbf{a}) \tag{6.2}$$

where $\xi_l(\mathbf{a}) \in \xi(\mathbf{a})$ is the fuzzy basis function vector and $\xi(\mathbf{a}) \in \boldsymbol{R}^M$, defined by $[\xi_1(\mathbf{a}), \ldots, \xi_M(\mathbf{a})]^T$. Further, $\xi_l(\mathbf{a})$ is calculated as

$$\xi_l(\mathbf{a}) = \frac{\prod_{p=1}^{m} \mu_{A_p^l}(a_p)}{\sum_{l=1}^{M} \left(\prod_{p=1}^{m} \mu_{A_p^l}(a_p) \right)} \tag{6.3}$$

In the fuzzy inference system, its knowledge base consists of fuzzy IF–THEN rules. These rules are empirical, which have to be established based on experts' knowledge about systems, controllers, performance, and so on. The input data of the fuzzy inference system are fuzzified by the fuzzifier. The fuzzy inference engine employs the rules and the fuzzified input data to achieve inference results. In order to apply the fuzzified inference results, the defuzzifier has to defuzzify the output data. Without a shadow of doubt, the introduction of input–output intervals and membership functions is more or less subjective, depending on the designer's experience and the available information.

The knowledge base plays a pivotal role in the performance of fuzzy inference system. Generally speaking, the more fuzzy rules the knowledge base contains, the better performance the fuzzy inference system has. However, this strategy yields an exponential growth in the number of rules as the input data. With the increasing of the input data in Figure 6.1, the fuzzy inference system suffers from the dilemma of exponential rule explosion.

The dilemma seriously hinders the applications of fuzzy control. The exponentially scaled rules result in the computational burden, which is one of the well-documented disadvantages of fuzzy control. Usually, an attempt is made to reduce the number of rules either by eliminating the number of rules that would not be encountered in the operating environment or in some way limiting the number of membership functions for each variable. But the attempt must decrease the performance of fuzzy control. Another solution to avoid the dilemma of rule explosion is to develop a novel fuzzy inference system.

6.2 Single-Input-Rule-Modules-Based Fuzzy Systems

The single-input-rule-modules (SIRM) model [3, 4] is a fuzzy inference system. In such a system, its antecedent part includes only one input item in each rule so that the number of rules is considerably reduced.

In a SIRM-based fuzzy inference system, only one input exists in each SIRM so that the increasing of fuzzy rules in its knowledge base can inherently avoid the exponential explosion. Since only one input exists in one SIRM, a fuzzy control system can contain several SIRM fuzzy inference systems. Without loss of generality, the ith SIRM of the fuzzy control system is taken as an example to illustrate the design of such a module. Define the ith SIRM as

$$\text{SIRM-}i : \{R_i^j : \ \text{IF } x_i \text{ is } A_i^j, \ \text{THEN } F_i \text{ is } C_i^j\}_{j=1}^{m_i} \tag{6.4}$$

Here SIRM-i is the ith SIRM of the fuzzy control system; R_i^j is the jth rule of SIRM-i; x_i and F_i are the antecedent and consequent variables of SIRM-i, respectively; F_i

is an intermediate variable corresponding to the output item F. A_i^j and C_i^j are the membership functions of x_i and F_i, respectively. Here $i = 1, 2, 3, \ldots, n$ denotes the index number of SIRMs, and $j = 1, 2, 3, \ldots, m_i$ denotes the rule number of SIRM-i.

For SIRM-i, its inference result F_i^0 of the consequent variable F_i can be calculated as

$$F_i^0 = \frac{\sum_{j=1}^{m_i} A_i^j(x_i)C_i^j}{\sum_{j=1}^{m_i} A_i^j(x_i)} \tag{6.5}$$

Intuitively, these SIRMs have a unique role in the performance of the fuzzy control system. Considering the system performance, some of the n SIRMs contribute more than others. In order to describe their roles and model the priority of the input data, a dynamic weight (DW) w_i^D is defined in eq. (6.6). Note that the weight does not depend on any input data:

$$w_i^D = w_i + B_i \Delta w_i^0 \tag{6.6}$$

In eq. (6.6), w_i and B_i are predefined parameters and w_i^0 is determined by fuzzy rules. Furthermore, w_i, a base value, guarantees that the ith input item plays the basic role; Δw_i^0, a dynamic variable, is automatically adjusted by the fuzzy rules according to the system performance. The ith input item in SIRM-i imposes on the system performance through its inference result F_i^0. Concerning the priority of the ith input item, F_i^0 in eq. (6.5) is dynamically weighted by w_i^D in eq. (6.6). Finally, the output item of all the n SIRMs F has the form

$$F = \sum_{i=1}^{n} w_i^D F_i^0 \tag{6.7}$$

6.3 SIRMs-Based Anti-sway Control Design

Consider the double-pendulum-type overhead crane system in Figure 2.2. Its control objective is to transport the payload to a target position as fast and accurately as possible; meanwhile, the crane control system should resist the double-pendulum oscillations by the control input, force f, applied to the trolley. In practice, the target position of the double-pendulum-type overhead crane system can be specified to any reasonable place. Without loss of generality, the positional range is limited to [−1 m 1 m], that is, $x \in$ [−1 m 1 m]. In order to measure the double-pendulum oscillations, a 0.5 rad angular threshold is empirically selected, that is, both θ_1 and θ_2 should be in the closed interval [−0.5 rad 0.5 rad] to guarantee the double-pendulum oscillations as small as possible. If either or both of θ_1 and θ_2 are above the threshold, it is accepted that the control action fails to suppress the double-pendulum oscillations.

The control objective contains three subobjectives: trolley positioning, hook stabilization, and payload stabilization. In order to simultaneously achieve these subobjectives, SIRMs-based fuzzy controller should cover three control components such as trolley-position control component, hook-angle control component, and payload-angle control component.

The double-pendulum-type overhead crane model (2.30) has three generable coordination variables. Assuming that these generable coordination variables and their first derivatives are measurable, all of them are adapted as input data such that the controller includes six input items. Furthermore, the driven force f is the sole control input in eq. (2.30) such that the controller includes only one output item. Concerning the design of the SIRMs, six SIRMs are allocated to the fuzzy controller so that each input item can contain a SIRM. To realize all the control components, these control components are hierarchical and their hierarchies are transient. To model the transient hierarchies, these SIRMs are dynamically weighted in light of their priority orders so that the controller can accomplish these control components in parallel. The output item of each SIRM is a part of the final control input.

To design SIRMs, the scaling factors of x, $\dot{x}$, θ_1, $\dot{\theta}_1$, θ_2, and $\dot{\theta}_2$ are fixed to 1.0 m, 1.0 m/s, 0.5 rad, 1.5 rad/s, 0.5 rad, and 1.5 rad/s. Here the maximum velocity and angular velocities are kept unknown and they are estimated to realize the SIRM design. The scaling factor of f is set up to $1.5(m_0 + m_1 + m_2)$ because f is affected by the total mass, where the coefficient is selected by means of trial and error. Having been normalized by their own private scaling factors, x, $\dot{x}$, θ_1, $\dot{\theta}_1$, θ_2, and $\dot{\theta}_2$ are assigned to input items $x_i (i = 1, 2, \ldots, 6)$ and f is assigned to the output item F. Note that x_i indicates the input item of SIRM-i and it does not denote the state variable.

In Chapter 5, the control design based on output feedback has been considered. Indeed, the assumption is not mild that x, $\dot{x}$, θ_1, $\dot{\theta}_1$, θ_2, and $\dot{\theta}_2$ are measurable for real industrial cranes. However, the assumption is mild enough for small-scale lab cranes because some sensors [5] and visual-based techniques [6] have been developed to measure angles. In order to focus on SIRM-based fuzzy control method for double-pendulum-type systems, we assume that x, $\dot{x}$, θ_1, $\dot{\theta}_1$, θ_2, and $\dot{\theta}_2$ are measurable even though sometimes measuring them is rather difficult in real industrial applications.

These SIRMs are inherently fuzzy inference systems determined by fuzzy rules. All the fuzzy rules have to be drawn from human experience because model (2.30) is assumed to be unavailable for control design. Although some results in Ref. [3] for a single-pendulum-type system are referential, how to design these SIRMs for the complex double-pendulum-type system (2.30) remains challenging. Here, these SIRMs are designed by analyzing the aforementioned three control components.

SIRMs of trolley-position control component. Two SIRMs such as SIRM-1 and SIRM-2 are allocated to realize the control component. Consider SIRM-1, whose antecedent and consequent variables are defined by x_1 and F_1, respectively. In order to achieve the aforementioned subobjective of trolley positioning, x_1 and F_1 should be opposite in direction according to Newton's second law. Meanwhile, x_1 and F_1 should be on the same tendency in magnitude for the purpose of fast positioning. Concerning SIRM-2, a similar conclusion can be drawn.

SIRMs of hook-angle and payload-angle control components. Not only have they coupled each other, but also both the control components are coupled with the trolley-position control component. The coupling also reflects the complexity of the double-pendulum-type system in Figure 2.2. Four SIRMs are allocated to realize the two control components. They are entitled SIRM-i ($i = 3, 4, 5, 6$). Correspondingly, x_i and F_i are adapted to describe the antecedent and consequent variables in SIRM-i. In order to design the four SIRMs, both the control components are mandatorily decoupled, indicating that the double-pendulum-type system is treated as two independent single-pendulum-type systems. Briefly, in order to stabilize the angles of the two independent single-pendulum systems, x_i and F_i in SIRM-i ($i = 3, 4, 5, 6$) should be united in the same direction; meanwhile, x_i and F_i should be on the same tendency in magnitude as well [3].

According to the aforementioned analyses, the fuzzy rules in SIRM-1 and SIRM-2 are described in Table 6.1. Furthermore, the fuzzy rules in other SIRMs are described in Table 6.2. In Tables 6.1 and 6.2, NB, ZO, and PB are linguistic labels and they denote negative big, zero, and positive big, respectively. The membership functions of these labels are defined in Figure 6.2.

Table 6.1: Fuzzy rules in SIRMs for trolley positioning

Antecedent variable $x_i (i = 1,2)$	**Consequent variable $F_i (i = 1,2)$**
NB	1.0
ZO	0.0
PB	−1.0

Table 6.2: Fuzzy rules in SIRMs for double-pendulum stabilization

Antecedent variable $x_i (i = 3,4,5,6)$	**Consequent variable $F_i (i = 3,4,5,6)$**
NB	−1.0
ZO	0.0
PB	1.0

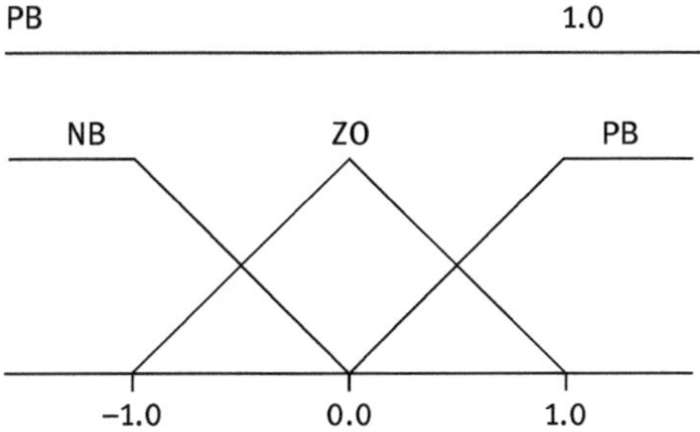

Figure 6.2: Membership functions of SIRM-1 and SIRM-2.

While designing SIRMs, the hook-angle and payload-angle control components are mandatorily decoupled. Let us provide some insights into such a design. Assuming that the double-pendulum-type crane system in Figure 2.2 is static, the hook and payload are in downward position. When the trolley moves toward the right direction by a positive-driven force, the hook will rotate clockwise; meanwhile, the payload will rotate counterclockwise. Apparently, the tendencies of the hook and payload are actually opposite even though the fuzzy rules in Table 6.2 are the same as each other. The contradiction between real tendencies and fuzzy rules will be dealing with DWs.

According to our experience, the three control components are transiently hierarchical. Such hierarchies indicate they have different degrees of importance. Although all SIRMs have been designed, these SIRMs cannot work until their DWs are designed. These DWs, modeling the transient hierarchies, are utilized to describe the priority orders of these SIRMs.

Formulated by eq. (6.6), these DWs are also determined by fuzzy logic so that their design is inherent to draw some fuzzy rules from human experience. Here, these DWs are designed by analyzing the motions of the double-pendulum-type system.

To solve the contradiction caused by the mandatory decoupling, four DWs of the last four SIRMs are considered in the first place, meaning the four SIRMs have higher priorities than the left two SIRMs. Consider the double-pendulum motions. In order to stabilize the hook and payload angles, one hypothesis is that the hook angle has higher priority than the payload angle. The hypothesis means that the hook-angle control component plays a more important role in eliminating the double-pendulum oscillations. Here some explanations about the hypothesis are given as follows.

Why does the hook angle have a higher priority than the payload angle? Figure 6.3 illustrates a dynamic double-pendulum-type system. In Figure 6.3(a), both the hook and payload slope are on the same side with respect to the vertical line. Moreover, θ_2 is larger than θ_1 but $\dot{\theta}_1$ and $\dot{\theta}_2$ are in opposite directions (displayed by dashed arches in

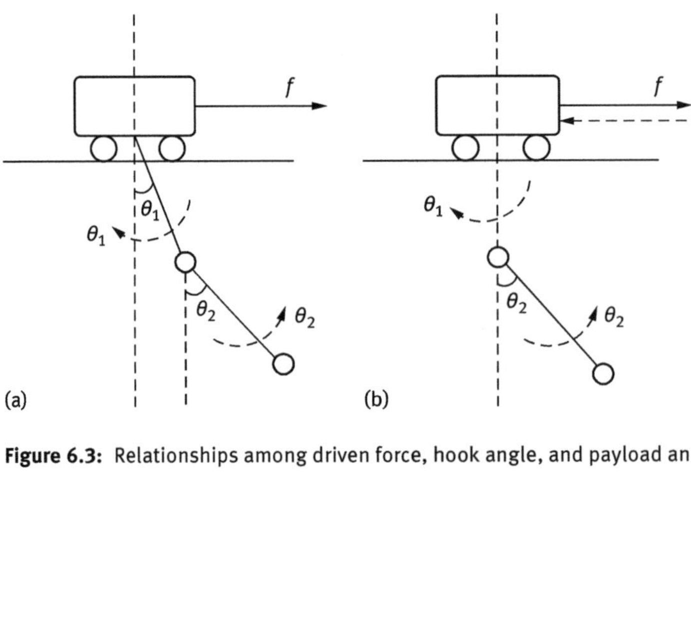

Figure 6.3: Relationships among driven force, hook angle, and payload angle.

Figure 6.3(a)). In this situation, stabilizing the double-pendulum-type system becomes more challenging because the mandatory decoupling results in contradictory fuzzy rules in Table 6.2. The contradiction can be dealt with based on human experience. Concerning θ_1 and $\dot{\theta}_1$, a control action probably fails when either or both of them are large. Furthermore, consider θ_2 and $\dot{\theta}_2$. The control action possibly works when either or both of them are large. Empirically, the hook angle has a higher priority than the payload angle so that a positive force needs to be applied to the trolley in Figure 6.3(a). Once the hook angle slopes on the opposite side as shown in Figure 6.3(b), the force needs to change its direction (dashed straight line in Figure 6.3(b)) regardless of the direction of the payload angle [7].

Take the DW of SIRM-i ($i = 3, 4, 5, 6$) into consideration. Based on the hypothesis, the absolute value of x_3, $|x_3|$, is selected as the antecedent variable in order to design the fuzzy rules. Correspondingly, Δw_i^0 ($i = 3, 4, 5, 6$) in eq. (6.6) is selected as the consequent variable.

Input items of SIRM-3 and SIRM-4 are concerned with the hook angle and angular velocity, respectively. For the purpose of strengthening them, $|x_3|$ and Δw_i^0 ($i = 3, 4$) should be on the same tendency in magnitude. Finally, fuzzy rules in DWs of SIRM-3 and SIRM-4 are determined in Table 6.3.

When the hook angle is small, we have empirical evidence that the double-pendulum-type system tends to be stable. In this situation, input items of SIRM-5 and SIRM-6 should be strengthened to accelerate the payload stabilization because they are concerned with the payload angle and angular velocity. Besides, they are not taken into consideration any more. Briefly, just when $|x_3|$ is small, Δw_i^0 ($i = 5, 6$) should be strengthened. Otherwise, $|x_3|$ has no effects on Δw_i^0 ($i = 5, 6$). Finally, fuzzy rules in DWs of SIRM-5 and SIRM-6 are determined in Table 6.4.

Table 6.3: Fuzzy rules in dynamic weights of SIRM-3 and SIRM-4

Antecedent variable $|x_3|$	**Consequent variables Δw_3^0 and Δw_4^0**
S	0.0
M	0.5
B	1.0

Table 6.4: Fuzzy rules in dynamic weights of SIRM-5 and SIRM-6

Antecedent variable $|x_3|$	**Consequent variables Δw_5^0 and Δw_6^0**
S	1.0
M	0.0
B	0.0

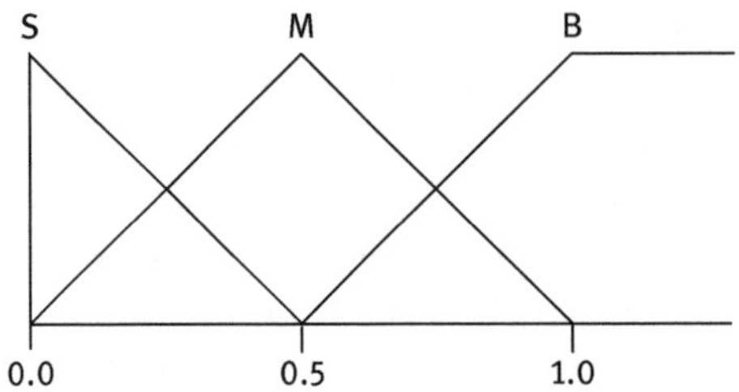

Figure 6.4: Membership functions SIRM-3, SIRM-4, SIRM-5 and SIRM-6.

In Tables 6.3 and 6.4, *S*, *M*, and *B* represent linguistic labels and they denote small, middle, and big, respectively. The membership functions of these linguistic labels are defined in Figure 6.4.

Take SIRM-1 and SIRM-2 into consideration. Although their input items are just concerned with the trolley position and velocity, their degrees of importance are not limited to trolley positioning. Recall the aforementioned transport control objective. The sole control input is utilized to realize the objective including three subobjectives. In other words, achieving these subobjectives by the sole control input results in tight coupling. Consequently, their DWs should indicate the tight coupling.

Concerning the DW of SIRM-1 and SIRM-2, the absolute values of x_3 and x_5, $|x_3|$ and $|x_5|$, are selected as the antecedent variable. Correspondingly, Δw_1^0 and Δw_2^0 are selected as the consequent variable. The fuzzy rules in both the DWs are also drawn from analyzing the motions of the double-pendulum-type system.

When either the hook angle or the payload angle is big, we have empirical evidence that the double-pendulum-type system tends to be unstable. In that case, the input items of SIRM-1 and SIRM-2 should not be strengthened so that SIRM-based control system can concentrate on the double-pendulum stabilization.

When either the hook angle or the payload angle is small, we have empirical evidence that the double-pendulum-type system tends to be stable. In that situation, input items of SIRM-1 and SIRM-2 should be strengthened in order to accelerate the trolley positioning.

Briefly, when either $|x_3|$ or $|x_5|$ is big, neither Δw_1^0 nor Δw_0^2 should be strengthened. When either $|x_3|$ or $|x_5|$ is small, both Δw_1^0 and Δw_2^0 should be strengthened. Finally, the fuzzy rules in the DWs of SIRM-1 and SIRM-2 are determined in Table 6.5. In Table 6.5, *S*, *M*, and *B* represent linguistic labels and they denote small, middle, and big, respectively. Their membership functions of the linguistic labels are defined in Figure 6.4.

From Table 6.4, it is concluded that if the hook angle is small, the payload-angle control component is strengthened by increasing the dynamic variables of SIRM-1 and SIRM-2. From Table 6.5, it is concluded that if either the hook angle or the payload angle is big, the trolley position control component is kept unchanged, and that if both the hook and payload angles are small, the trolley-position control component is strengthened by increasing the dynamic variables of SIRM-1 and SIRM-2.

Table 6.5: Fuzzy rules in dynamic weights of SIRM-1 and SIRM-2

Consequent variables Δw_5^0 and Δw_6^0		**Antecedent variable $|x_3|$**		
		S	*M*	*B*
Antecedent variable $|x_5|$	*S*	1.0	0.5	0.0
	M	0.5	0.0	0.0
	B	0.0	0.0	0.0

Control system structure. Consider the double-pendulum-type system in Figure 2.2. The structure of its SIRM-based fuzzy control system is presented in Figure 6.5. From Figure 6.5, x, $\dot{x}$, θ_1, $\dot{\theta}_1$, θ_2, and $\dot{\theta}_2$ are fed back. Their errors with respect to their reference values are normalized via the normalizer (NL) block. Then, the fuzzy input items x_i $(i = 1, 2, \ldots, 6)$ are obtained. The input item x_i feeds to the SIRM-i block and the output of SIRM-i, F_i^0, is obtained. Both x_3 and x_5 feed the two DW blocks of the trolley as the antecedent variables. The two DW blocks of the hook employ the input item x_3 as their antecedent variable, so do the payload DW blocks. Summing up each output of the SIRM-i block, F_i^0, times the output of its own DW block, w_i^D, the output item F of the SIRM-based fuzzy controller can be obtained. Inputting f to the output scaling factor block, the control input, that is, driven force f, can be finally drawn.

6.4 Simulation Results

The SIRM-based fuzzy controller will be applied to the transport control problem of double-pendulum-type overhead crane systems. Concerning the double-pendulum-type system in Figure 2.2, some physical parameters are determined by $m_0 = 0.5$ kg, $m_1 = 0.25$ kg, $m_2 = 0.5$ kg, $l_1 = 0.5$ m, and $l_2 = 0.5$ m. The parameters are regarded as nominal parameters. The scaling factors of the input items have been defined during the control design as well as the scaling factor of the output item. Although all of the mass and length values are small when compared with a real crane, they can be scaled up in practice. On the other hand, parameters are available for a small lab-scaled testbed.

$w_i^D (i = 1, 2, \ldots, 6)$ in eq. (6.6) has three parameters to be determined as well. They are Δw_i^0, w_i, and B_i, respectively, where Δw_i^0 has been discussed and it is decided by fuzzy rules. The other two parameters need to be predefined. They are listed in Table 6.6 and are picked up by trial and error.

To verify the feasibility and validity of the SIRM-based fuzzy controller for the transport control problem, numerical simulations are carried out. The control objective is to transport the payload to a desired position. Without loss of generality, the desired position x_d is located at 1 m, that is, $x_d = 1$ m from the origin in Figure 2.2. Meanwhile, the controller needs to suppress the double-pendulum motions of the hook and payload. The results are presented in Figures 6.6–6.12.

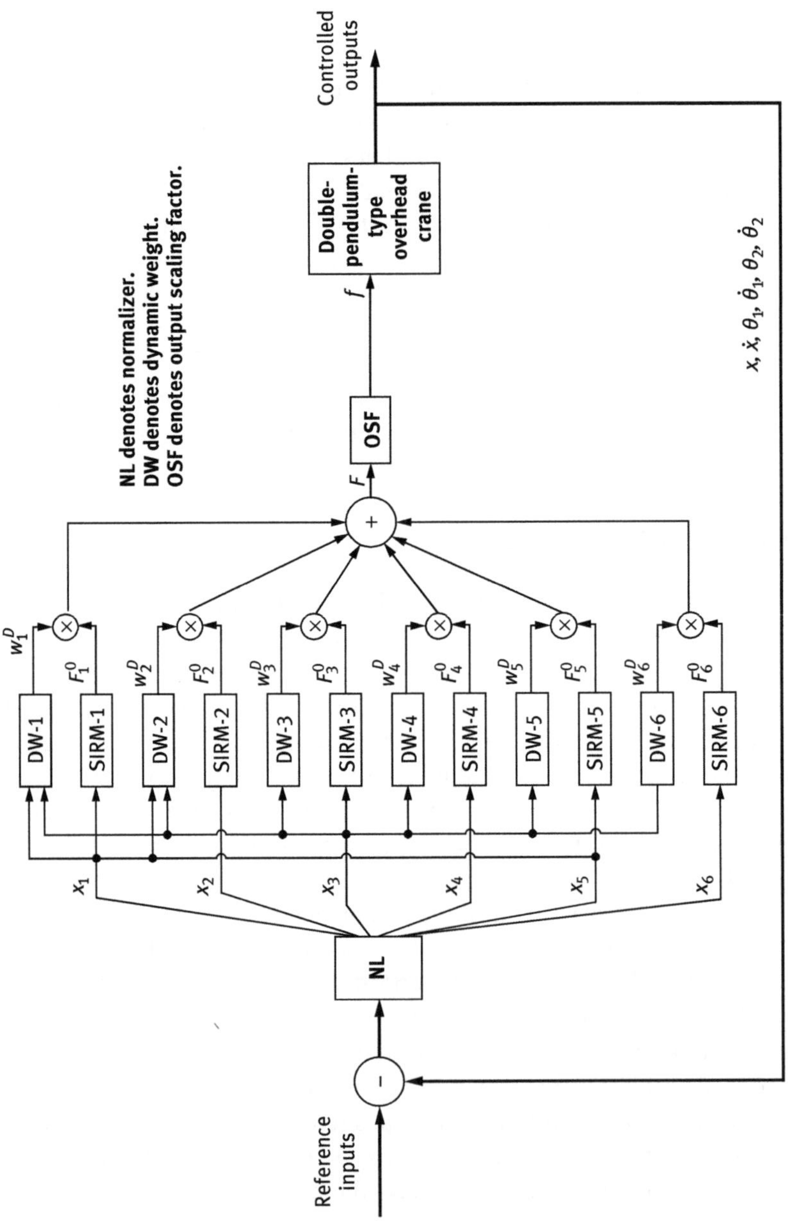

Figure 6.5: Block diagram of the SIRM-based fuzzy control system.

Table 6.6: Predefined parameters in all dynamic weights

Input item x_i	Base value w_i	Breadth B_i
x_1	1.3015	2.2103
x_2	1.3817	2.1729
x_3	0.6970	0.3151
x_4	0.7018	0.2852
x_5	0.0006	0.2028
x_6	0.0003	0.1946

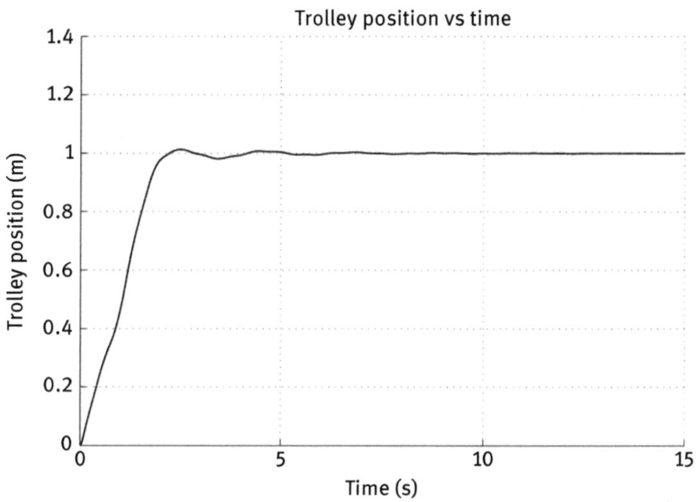

Figure 6.6: Trolley position versus time by the SIRM-based fuzzy control.

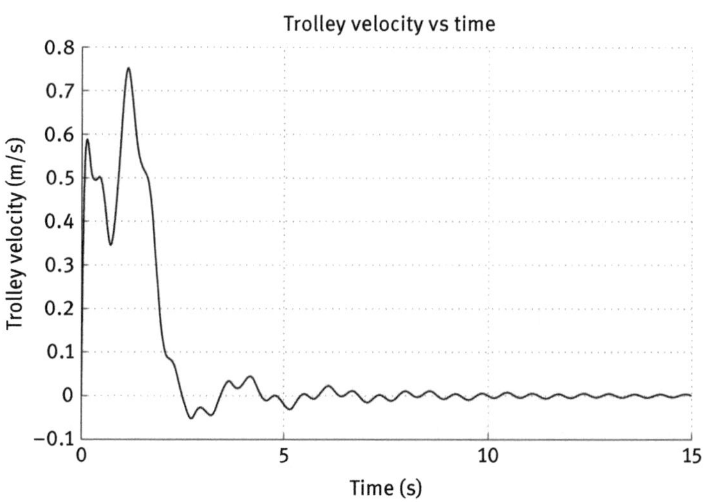

Figure 6.7: Trolley velocity versus time by the SIRM-based fuzzy control.

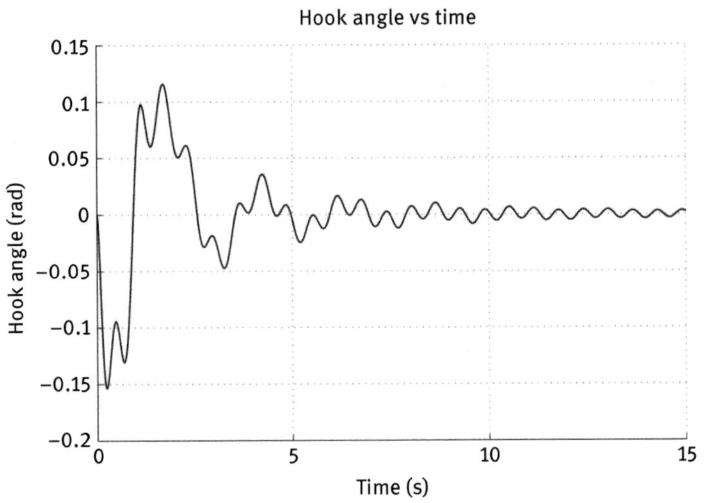

Figure 6.8: Hook angle versus time by the SIRM-based fuzzy control.

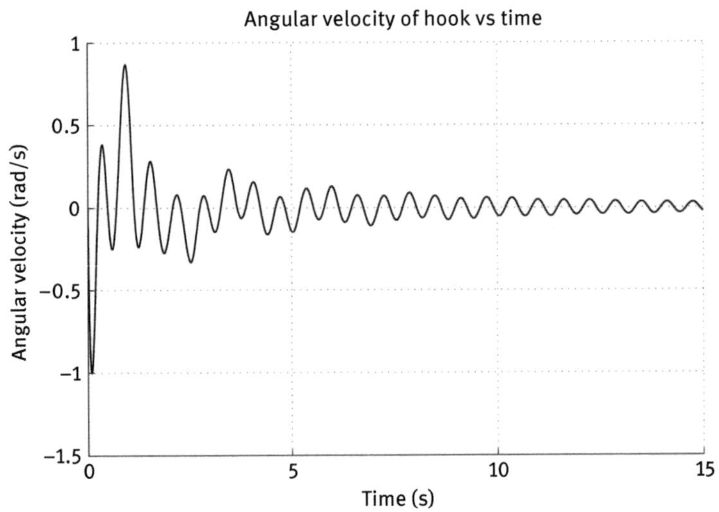

Figure 6.9: Angular velocity of hook versus time by the SIRM-based fuzzy control.

Displayed in Figures 6.6–6.12, the SIRM-based fuzzy controller is able to achieve the transport control objective. Although the curve in Figure 6.6 arrives at the desired position at about 6 s, the curves in Figures 6.8 and 6.10 illustrate that the two angular deviations are not eradicated until 10 s. In order to keep the two angular deviations as small as possible, the driven force f in Figure 6.12 sharply changes at the outset. When the trolley is close to its desired position, f is mildly applied to the double-pendulum-type system and changes the direction according to the designed fuzzy rules till the

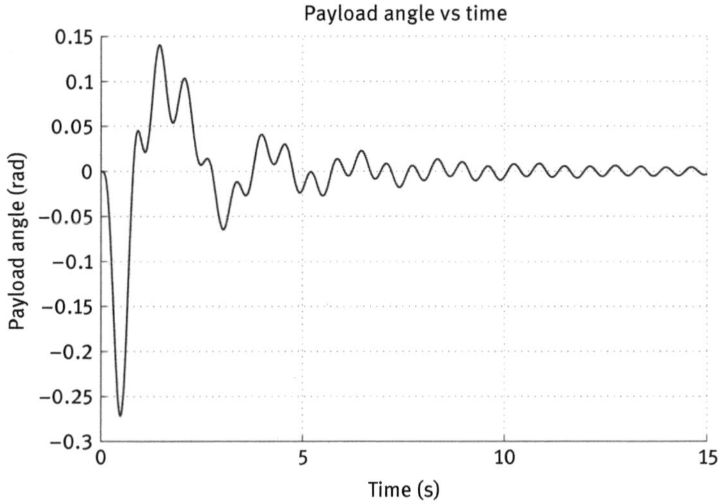

Figure 6.10: Payload angle versus time by the SIRM-based fuzzy control.

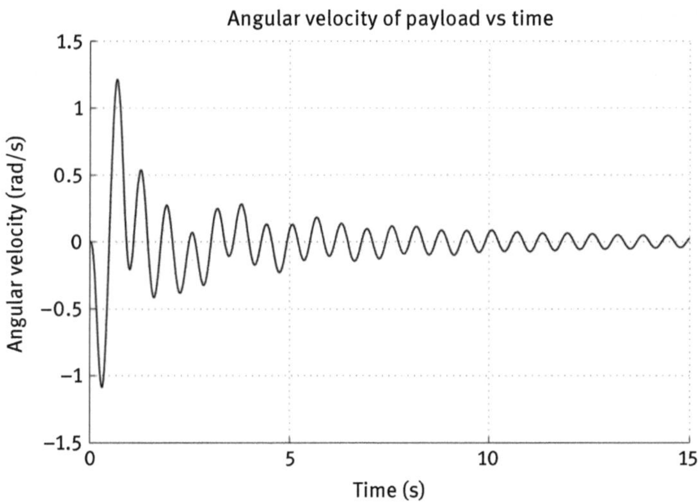

Figure 6.11: Angular velocity of payload versus time by the SIRM-based fuzzy control.

two angular deviations tend to zero. Displayed in Figures 6.8–6.11, the hook and payload angles still have some tiny residual oscillations when the trolley arrives at the desired position. From the control design, one advantage of the SIRM-based fuzzy control method is that the method is completely free of the crane model. All fuzzy rules are empirically drawn from intuition and experience.

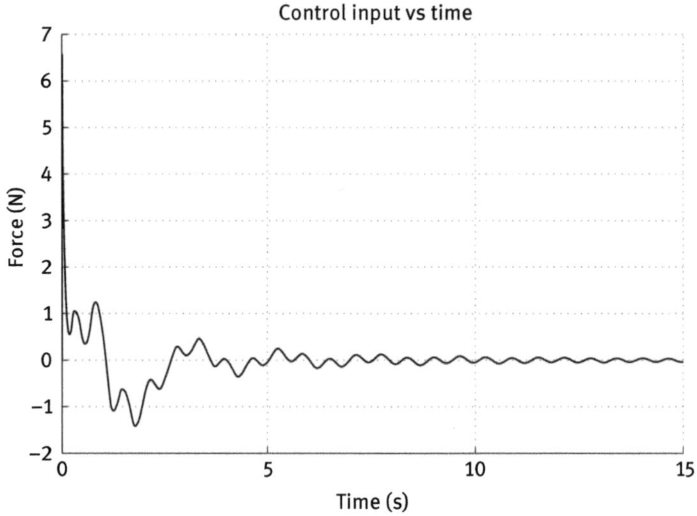

Figure 6.12: Control input versus time by the SIRM-based fuzzy control.

6.5 Source Codes

Figures 6.6–6.12 can be numerically obtained by Matlab software [8]. The source codes are listed as follows:

Matlab codes to plot Figures 6.6–6.12

```
function SIRM
clear all;
close all;

len1=0.5;
len2=0.5;
mass=0.5;
mass1=0.25;
mass2=0.5;
g=9.81;

s=0;
v=0;
th1=0;
```

```
th2=0;
w1=0;
w2=0;
f=0;
t=0;
s_1=s;
v_1=v;
th_1=th1;
th_2=th2;
w_1=w1;
w_2=w2;
dt=0.01;

xend=1;

sdata=s;
vdata=v;
th1data=th1;
th2data=th2;
w1data=w1;
w2data=w2;
fdata=f;
tdata=t;

for i=1:1:1500
s=s_1;
v=v_1;
th1=th_1;
th2=th_2;
w1=w_1;
w2=w_2;

M=[mass+mass1+mass2 (mass1+mass2)*len1*cos(th1)
   mass2*len2*cos(th2);...
```

```
        (mass1+mass2)*len1*cos(th1) (mass1+mass2)*len1^2
            mass2*len1*len2*cos(th1-th2);...
        mass2*len2*cos(th2) mass2*len1*len2*cos(th1-th2)
            mass2*len2^2];

C=[0 -1*(mass1+mass2)*len1*w1*sin(th1)
  -1*mass2*len2*w2*sin(th2);...
      0 0 mass2*len1*len2*w1*sin(th1-th2);...
      0 -1*mass2*len1*len2*w1*sin(th1-th2) 0];
G=[0;(mass1+mass2)*g*len1*sin(th1);mass2*g*len2*sin(th2)];

f=controller(mass,mass1,mass2,xend,s,v,th1,th2,w1,w2);
T=[f;0;0];
x=inv(M)*(T-G-C*[v;w1;w2]);
y=[v;w1;w2]+dt*x;
v_1=y(1);
w_1=y(2);
w_2=y(3);
z=[s;th1;th2]+0.5*dt^2*x+dt*[v;w1;w2];
s_1=z(1);
th_1=z(2);
th_2=z(3);
t=t+dt;
tdata = [tdata,t];
fdata = [fdata,f];
vdata = [vdata,v_1];
th1data=[th1data,th_1];
th2data=[th2data,th_2];
w1data=[w1data,w_1];
w2data=[w2data,w_2];
sdata=[sdata,s_1];
end

h1=figure(1);
```

```
set(h1,'Position',[100,200,500,300],...
      'Menu','none',...
      'NumberTitle','off',...
      'Name','Trolley position'...
);
axes1=axes('Parent',h1);
title(axes1,'Trolley position vs time','FontSize',16);
xlabel(axes1,'Time(s)');
ylabel(axes1,'Trolley position(m)');
hold(axes1,'all');
plot(tdata,sdata);
grid on;

h1=figure(2);
set(h1,'Position',[100,200,500,300],...
      'Menu','none',...
      'NumberTitle','off',...
      'Name','Trolley velocity'...
);
axes1=axes('Parent',h1);
title(axes1,'Trolley velocity vs time','FontSize',16);
xlabel(axes1,'Time(s)');
ylabel(axes1,'Trolley velocity(m/s)');
hold(axes1,'all');
plot(tdata,vdata);
grid on;

h2=figure(3);
set(h2,'Position',[800,200,500,300],...
      'Menu','none',...
      'NumberTitle','off',...
      'Name','Hook angle'...
);
axes2=axes('Parent',h2);
```

```
title(axes2,'Hook angle vs time','FontSize',16);
xlabel(axes2,'Time(s)');
ylabel(axes2,'Hook angle(rad)');
hold(axes2,'all');
plot(tdata,th1data,'r');
grid on;

h2=figure(4);
set(h2,'Position',[800,200,500,300],...
     'Menu','none',...
     'NumberTitle','off',...
     'Name','Angular velocity'...
);
axes2=axes('Parent',h2);
title(axes2,'Angular velocity of hook vs time','FontSize',16);
xlabel(axes2,'Time(s)');
ylabel(axes2,'Angular velocity(rad/s)');
hold(axes2,'all');
plot(tdata,w1data,'r');
grid on;

h2=figure(5);
set(h2,'Position',[800,200,500,300],...
     'Menu','none',...
     'NumberTitle','off',...
     'Name','Payload angle'...
);
axes2=axes('Parent',h2);
title(axes2,'Payload angle vs time','FontSize',16);
xlabel(axes2,'Time(s)');
ylabel(axes2,'Payload angle(rad)');
hold(axes2,'all');
plot(tdata,th2data,'r');
grid on;
```

```
h2=figure(6);
set(h2,'Position',[800,200,500,300],...
      'Menu','none',...
      'NumberTitle','off',...
      'Name','Angular velocity'...
);
axes2=axes('Parent',h2);
title(axes2,'Angular velocity of payload vs time','FontSize',16);
xlabel(axes2,'Time(s)');
ylabel(axes2,'Angular velocity(rad/s)');
hold(axes2,'all');
plot(tdata,w2data,'r');
grid on;

h2=figure(7);
set(h2,'Position',[800,200,500,300],...
      'Menu','none',...
      'NumberTitle','off',...
      'Name','Control input'...
);
axes2=axes('Parent',h2);
title(axes2,'Control input vs time','FontSize',16);
xlabel(axes2,'Time(s)');
ylabel(axes2,'Force(N)');
hold(axes2,'all');
plot(tdata,fdata,'r');
grid on;

function f = controller(mass,mass1,mass2,xend,s,v,th1,th2,w1,w2 )

a1=th1/0.5;
b1=w1/1.5;
if a1>-1&&a1<0
   u1=(a1^2+2*a1)/(-2*a1^2-2*a1+1);
```

```
elseif a1>0&&a1<1
   u1=(2*a1-a1^2)/(-2*a1^2+2*a1+1);
elseif a1<-1
   u1=-1;
elseif a1==0
   u1=0;
elseif a1>1
   u1=1;
end
if b1>-1&&b1<0
   u2=(b1^2+2*b1)/(-2*b1^2-2*b1+1);
elseif b1>0&&b1<1
   u2=(2*b1-b1^2)/(-2*b1^2+2*b1+1);
elseif b1<-1
   u2=-1;
elseif b1==0
   u2=0;
elseif b1>1
   u2=1;
end
a2=th2/0.5;
b2=w2/1.5;
if a2>-1&&a2<0
   u3=(a2^2+2*a2)/(-2*a2^2-2*a2+1);
elseif a2>0&&a2<1
   u3=(2*a2-a2^2)/(-2*a2^2+2*a2+1);
elseif a2<-1
   u3=-1;
elseif a2==0
   u3=0;
elseif a2>1
   u3=1;
end
if b2>-1&&b2<0
```

```
    u4=(b2^2+2*b2)/(-2*b2^2-2*b2+1);
elseif b2>0&&b2<1
    u4=(2*b2-b2^2)/(-2*b2^2+2*b2+1);
elseif b2<-1
    u4=-1;
elseif b2==0
    u4=0;
elseif b2>1
    u4=1;
end
if (xend-s)>=-1&&(xend-s)<0
    u5=((xend-s)^2+2*(xend-s))/(-2*(xend-s)^2-2*(xend-s)+1);
elseif (xend-s)>0&&(xend-s)<=1
    u5=(2*(xend-s)-(xend-s)^2)/(-2*(xend-s)^2+2*(xend-s)+1);
elseif (xend-s)<-1
    u5=-1;
elseif (xend-s)==0
    u5=0;
elseif (xend-s)>1
    u5=1;
end
if v>-1&&v<0
    u6=(-1*v^2-2*v)/(-2*v^2-2*v+1);
elseif v>0&&v<1
    u6=(v^2-2*v)/(-2*v^2+2*v+1);
elseif v<-1
    u6=1;
elseif v==0
    u6=0;
elseif v>1
    u6=-1;
end
if th1>-0.5&&th1<0.5
    r1=(abs(th1)-abs(th1)^2)/(0.5+2*abs(th1)-4*abs(th1)^2);
```

```
   r2=1;
elseif abs(th1)>=0.5&&abs(th1)<=1
   r1=(5*abs(th1)-3*abs(th1)^2-1.5)/(6*abs(th1)-4*abs(th1)
        ^2-1.5);
   r2=0;
elseif abs(th1)>1
   r1=1;
   r2=0;
end

   a3=max(abs(a1),abs(a2));
   b3=min(abs(a1),abs(a2));

if abs(a1)>=0.5&&abs(a1)<=1&&abs(a2)>=0.5&&abs(a2)<=1
   r3=0;
elseif a1>-0.5&&a1<0.5&&a2>-0.5&&a2<0.5
   r3=(a3-a3^2+0.5-2*b3^2)/(2*a3-2*a3^2+0.5-2*b3^2);
elseif a1>-0.5&&a1<0.5&&abs(a2)>=0.5&&abs(a2)<=1
   if (0.5-abs(a1))<=(abs(a2)-0.5)
      r3=(abs(a1)-3*a1^2+0.5)/(4*abs(a2)
         -2*a2^2-4*a1^2+2*abs(a1)-1);
   elseif (0.5-abs(a1))>(abs(a2)-0.5)
      r3=(0.5-2*a1^2+abs(a2)-a2^2)/(6*abs(a2)
         -4*a2^2-2*a1^2-1);
   end
elseif a2>-0.5&&a2<0.5&&abs(a1)>=0.5&&abs(a1)<=1
   if (0.5-abs(a2))<=(abs(a1)-0.5)
      r3=(abs(a2)-3*a2^2+0.5)/(4*abs(a1)
         -2*a1^2-4*a2^2+2*abs(a2)-1);
   elseif (0.5-abs(a2))>(abs(a1)-0.5)
      r3=(0.5-2*a2^2+abs(a1)-a1^2)/(6*abs(a1)
         -4*a1^2-2*a2^2-1);
   end
elseif abs(a1)>1||abs(a2)>1
```

```
    r3=0;
  end
  f=1.5*(mass+mass1+mass2)*((0.7+0.3*r1)*(u1+u2)+(0.2*r2)*
       (u3+u4)+(1.3+2.2*r3)*(u5+u6));
```

References

[1] Chen G, Pham TT. Introduction to fuzzy sets, fuzzy logic, and fuzzy control systems. Padstow: CRC Press, 2000.

[2] Yager RR, Zadeh LA. An introduction to fuzzy logic applications in intelligent systems. Berlin: Springer, 2012.

[3] Yi J, Yubazaki N, Hirota K. Anti-swing and positioning control of overhead traveling crane. Inf Sci, 2003, 155(1): 19–42.

[4] Seki H, Ishii H, Mizumoto M. On the generalization of single input rule modules connected type fuzzy reasoning method. IEEE Trans Fuzzy Syst, 2008, 16(5): 1180–1187.

[5] Liu D, Yi J, Zhao D, Wang W. Adaptive sliding mode fuzzy control for a two-dimensional overhead crane. Mechatronics, 2005, 15(5): 505–522.

[6] Lee LH, Huang CH, Ku SC, Yang ZH, Chang CY. Efficient visual feedback method to control a three-dimensional overhead crane. IEEE Trans Ind Electron, 2014, 61(8): 4073–4083.

[7] Qian DW, Tong SW, Lee S. Fuzzy-logic-based control of payloads subjected to double-pendulum motion in overhead cranes. Autom Constr, 2016, 65: 133–143.

[8] Xue D, Chen Y, Atherton DP. Linear feedback control: Analysis and design with MATLAB. Philadelphia, PA: Society for Industrial and Applied Mathematics, 2007.

7 Input-Shaping-Based Anti-sway Control

Abstract: This chapter investigates the methodology of input shaping for the anti-sway control problem of double-pendulum-type overhead cranes. Input shaping is an open-loop control technology. It can implement its control action by intelligently shaping the reference commands. The method can effectively cancel motion-induced oscillations. Inherently, the method is a feed-forward technique. In real applications, it is hard to resist any error without special design. In order to deal with these issues, this chapter starts with some basic concepts about input shaping. One simultaneous input shaper is designed to overcome the double-pendulum oscillations. Meanwhile, one classic convolved input shaper is illustrated as a comparison. Then, the simultaneous input shaper integrated with a single-input-rule module-based fuzzy controller is explored for the transport control problem of double-pendulum-type overhead cranes. Some simulation results are illustrated to support the hybrid control design. Finally, all the source codes are listed.

Keywords: Simultaneous input shaping, convolved input shaping, single-input-rule modules, double-pendulum-type motions, overhead cranes

7.1 Input Shaping

Input shaping is a control technique for reducing vibrations [1]. The method works by creating a command signal, where vibration caused by the first part of the command signal is canceled by vibration caused by the second part of the command signal. Input shaping can be considered as a predictive control scheme. That is, it uses knowledge of the system to reshape reference commands so that undesired system behavior does not occur. Therefore, input shaping allows for quick, low-vibration motions. The input-shaping method has proven effective for controlling systems with oscillations, such as cranes and flexible structures.

Input shaping has many advantages and can benefit us a lot [2]. The method is a much different requirement than knowing the states of the system in real time, as is needed with feedback control. To design an input shaper, these estimates can be determined, and the design can be completed off-line. Input shaping does not require cameras or other sensors to determine the system states so that it is less expensive to implement. Otherwise, when using input shaping, the system can have a faster settling time than with feedback controllers alone. Once the shaper is designed, no further knowledge of the system states is needed.

Take the input-shaping process of a system as an example [3]. A series of impulses is convolved in real time with the original reference command to create a shaped command. The original velocity pulse command is transformed into a staircase command that moves the system without oscillation. The amplitudes and time locations of the

DOI 10.1515/9783110520101-007

impulses that compose the input shaper are designed using estimates of the natural frequencies and damping ratios. There are always errors in these estimates, so the input shaper has to be made robust to changes in these parameters. Unfortunately, the input-shaping method is a traditional open-loop control strategy. The effectiveness of traditional input shaping can be degraded when the unexpected occurs, such as modeling errors, disturbances, nonlinearities, and so on. In order to determine the impulse amplitudes and time locations of an input shaper, one needs to ensure that the amplitudes and locations satisfy certain design constraints such that the input shaper can work well under constraints. Some typical constraints [4] can be considered as follows:

- The primary design constraint is a limit on the amplitude of vibration caused by the shaper, formulated as

$$V_{\text{amp}} \leq V_{\text{tol}} \tag{7.1}$$

  Here V_{amp} yields the percentage residual vibration and V_{tol} denotes a tolerable level of residual vibration.
- To ensure the fastest solution possible, the time of the last impulse is typically minimized, having the form

$$\min(t_n) \tag{7.2}$$

  Here t_n is the nth impulse in a sequence composed of n-impulses.
- Impulse amplitudes are also limited to sum to 1 in order that the shaped command can reach the same set point as the unshaped command, determined as

$$\sum_{i=1}^{n} A_i = 1 \tag{7.3}$$

  where A_i means the ith impulse amplitude.
- The impulse amplitudes will be driven toward positive and negative infinity.

One impulse amplitude constraint is to limit impulses to have positive amplitude. This indicates that positive shapers have impulse amplitudes between zero and one, that is,

$$0 < A_i \leq 1 \tag{7.4}$$

For unity magnitude input shapers, the impulse amplitudes are constrained to be ±1. This constraint can be stated as

$$\begin{aligned} &0 < A_i \leq 1 \text{ when } i \text{ is odd} \\ &A_i = -A_{\max} \text{ when } i \text{ is even} \end{aligned} \tag{7.5}$$

Concerning the areas of modeling errors and nonlinearities, a lot of research [5, 6] has been done to make the input-shaping method more robust. However, as with an open-loop technique, the input-shaping method can do nothing to address disturbances. To deal with the issue, the intelligent combination of input shaping and feedback control is a potential solution.

7.2 Input-Shaping-Based Anti-sway Control Design

7.2.1 Control Structure

Concerning the combination of input shaping and feedback control, the control structure should be defined at first. Generally speaking, there are two kinds of structures [1]. One is outside-the-loop input shaping, meaning that the input shaper is placed outside the feedback loop as shown in Figure 7.1. The other is inside-the-loop input shaping, meaning that the input shaper is placed somewhere inside the feedback loop as shown in Figure 7.2.

From the block diagrams, the shaper in Figure 7.1 is employed to shape the reference signal, which acts as a shaper. On the other hand, the shaper in Figure 7.2 is located in front of the controller and it acts as a filter to filter the control signal generated by the controller. The filtered control signal is implemented on the plant. Note that the structure in Figure 7.1 is the traditional form of input shaping. Meanwhile, the shaper in Figure 7.2 can be designed in the same manner as a standard input shaper even if it acts as a filter.

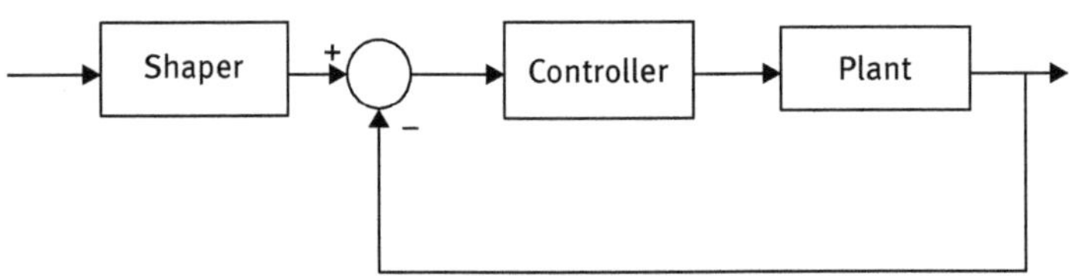

Figure 7.1: Block diagram of outside-the-loop input shaping.

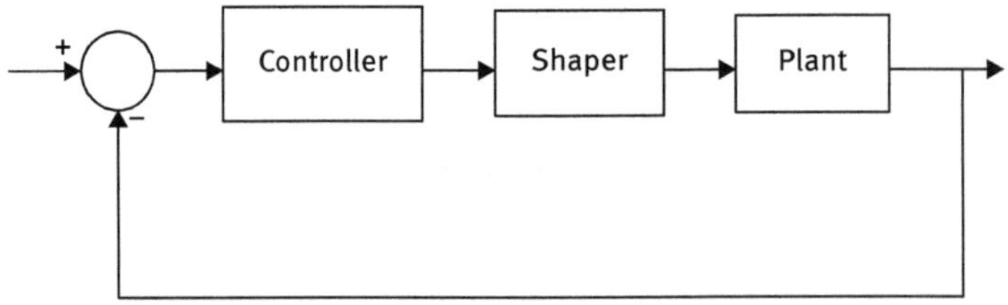

Figure 7.2: Block diagram of inside-the-loop input shaping.

In this chapter, the motivation is to give some insights into the combination of input shaping and the aforementioned control methods for double-pendulum-type overhead cranes. Without loss of generality, the structure in Figure 7.2 is adopted in the following design.

7.2.2 Control Design

7.2.2.1 Convolved Input-Shaping-Based Anti-sway Control Design

Since the primary motivation of input shaping is to reduce oscillations, the zero vibration (ZV) shaper is developed to limit the residual vibration to zero at the design frequency [4]. Such a shaper completely depends on system parameters. If system parameters are not exactly known, the shaper performance may be dramatically decreased. In reality, modeling system parameters within the tolerance needed for ZV shapers is rather difficult, and sometimes impossible. This challenge inspired the development of robust shapers.

In order to increase the robustness of the ZV shaper, the derivative method is taken into consideration, that is, setting the derivative, with respect to the frequency of the residual vibration to zero. Such an improved ZV shaper is called a zero vibration and derivative (ZVD) shaper [4].

There are numerous ways to design single-mode input shapers [7]. However, recall Chapter 2, where eqs. (2.49) and (2.50) indicate that the double-pendulum-type overhead crane system in Figure 2.2 has two linearized frequencies. In order to suppress oscillations with two modes, the convolved ZVD input shaper $\mathrm{ZVD_{cov}}$ has to be developed and to suppress the oscillation with a frequency f_0, its ZVD shaper ZVD_{f0} can be designed as follows:

$$\mathrm{ZVD}_{f_0} = \begin{bmatrix} A_i \\ t_i \end{bmatrix} = \begin{bmatrix} \frac{1}{4} & \frac{1}{2} & \frac{1}{4} \\ 0 & \frac{\pi}{f_0} & \frac{2\pi}{f_0} \end{bmatrix} \tag{7.6}$$

Concerning the double-pendulum motions in Figure 2.2, oscillations contain two frequencies, that is, the high frequency ω_1 and the low frequency ω_2. The two ZVD shapers with respect to two frequencies can be determined by

$$\begin{aligned} \mathrm{ZVD}_{\omega_1} &= \begin{bmatrix} A_i \\ t_i \end{bmatrix} = \begin{bmatrix} \frac{1}{4} & \frac{1}{2} & \frac{1}{4} \\ 0 & \frac{\pi}{\omega_1} & \frac{2\pi}{\omega_1} \end{bmatrix} \\ \mathrm{ZVD}_{\omega_2} &= \begin{bmatrix} A_i \\ t_i \end{bmatrix} = \begin{bmatrix} \frac{1}{4} & \frac{1}{2} & \frac{1}{4} \\ 0 & \frac{\pi}{\omega_2} & \frac{2\pi}{\omega_2} \end{bmatrix} \end{aligned} \tag{7.7}$$

Convolving the two shapers [8] yields

$$\mathrm{ZVD_{cov}} = \begin{bmatrix} A_i \\ t_i \end{bmatrix} = \begin{bmatrix} \frac{1}{16} & \frac{1}{8} & \frac{1}{16} & \frac{1}{8} & \frac{1}{4} & \frac{1}{8} & \frac{1}{16} & \frac{1}{8} & \frac{1}{16} \\ 0 & \frac{\pi}{\omega_1} & 2\frac{\pi}{\omega_1} & \frac{\pi}{\omega_2} & \frac{\pi}{\omega_1} + \frac{\pi}{\omega_2} & 2\frac{\pi}{\omega_1} + \frac{\pi}{\omega_2} & 2\frac{\pi}{\omega_2} & \frac{\pi}{\omega_1} + 2\frac{\pi}{\omega_2} & 2\frac{\pi}{\omega_1} + 2\frac{\pi}{\omega_2} \end{bmatrix} \tag{7.8}$$

Equation (7.8) is just the convolved ZVD input shaper ZVD_{cov}. Finally, the force applied to the double-pendulum-type crane in Figure 2.2 is generated by convolving the feedback control law u with the convolved shaper:

$$f = u * ZVD_{cov} \tag{7.9}$$

Here * denotes the convolving operator.

7.2.2.2 Simultaneous Input-Shaping-Based Anti-sway Control Design

The basic idea of the simultaneous input shaper is to make the shaper duration t_n in eq. (7.2) as short as possible. Technically, the simultaneous input shaper design is to minimize the time of the input shaper impulse subjected to some constraints, where the objective function or penalty function is formulated as

$$\text{objective function} \quad \min t_n \tag{7.10}$$

The following constraints need to be considered in the optimization problem:

$$\text{Residual vibration} \quad V_{amp} \le V_{tol} \tag{7.11}$$

In eq. (7.11), $V_{amp} = |C_1| + |C_2|$, where C_1 and C_2 are defined by eqs. (2.56) and (2.57):

$$\text{Impulse amplitudes} \sum_{i=1}^{n} A_i = 1 \tag{7.12}$$

Since the positive impulses are adapted, we have

$$\text{Positive shaper} \quad A_i > 0 \tag{7.13}$$

The aforementioned constraints can be drawn from the concept of input shaping. In addition, a five-impulse shaper is employed to suppress the double-pendulum motions so that one extra constraint is listed as

$$\text{Symmetry} \quad \begin{matrix} A_1 - A_5 = 0 \\ A_2 - A_4 = 0 \end{matrix} \tag{7.14}$$

The solution of the optimization problem can be dealt with the MATLAB optimization toolbox. Consider the double-pendulum-type overhead crane in Figure 2.2. Its physical parameters are determined as $m_0 = 0.5$ kg, $m_1 = 0.25$ kg, $m_2 = 0.5$ kg, $l_1 = 0.5$ m, $l_2 = 0.5$ m, and $g = 9.81$ m/s^2. After four iterations, the solution of the designed simultaneous input shaper SIS_{shaper} is obtained as

$$SIS_{shaper} = \begin{bmatrix} A_i \\ t_i \end{bmatrix} = \begin{bmatrix} 0.2152 & 0.2848 & 0.0000 & 0.2848 & 0.2152 \\ 0 & 0.2382 & 0.8727 & 0.9908 & 1.2290 \end{bmatrix} \tag{7.15}$$

Equation (7.15) is just the simultaneous input shaper. Finally, the force applied to the double-pendulum-type crane in Figure 2.2 is generated by convolving the feedback control law u with the shaper [8]:

$$f = u^{*}\mathrm{SIS}_{\mathrm{shaper}} \tag{7.16}$$

Here * denotes the convolving operator.

7.3 Simulation Results

With regard to the control structure in Figure 7.2, any feedback control law can integrate the convolved input shaper $\mathrm{ZVD}_{\mathrm{cov}}$ and the simultaneous input shaper $\mathrm{SIS}_{\mathrm{shaper}}$.

7.3.1 Convolved Input-Shaping-Based Anti-sway Control

Adapt the structure in Figure 7.2 and integrate the convolved input-shaping-based anti-sway control and the passivity-based control. Such a control design is presented to realize the transport control of double-pendulum-type overhead crane systems.

Physical parameters of the double-pendulum-type crane system are determined as $m = 0.5$ kg, $m_1 = 0.25$ kg, $m_2 = 0.5$ kg, $l_1 = 0.5$ m, and $l_2 = 0.5$ m.

The passivity-based controller [9] has been proposed in Chapter 3. It has been implemented on the crane system with the same physical parameters. The convolved input shaper has been designed in eq. (7.8). Finally, the control law applied to the crane system is determined by eq. (7.9).

The control objective is to transport the payload to a desired position. Without loss of generality, the desired position x_d is located at 1 m, that is, $x_d = 1$ m from the origin in Figure 2.2. Meanwhile, the controller needs to suppress the double-pendulum motions of the hook and payload. The results are presented in Figures 7.3–7.9.

As displayed in Figures 7.3–7.9, the inside-the-loop design that integrates the controller and the shaper is able to achieve the transport control objective. Although the curve in Figure 7.3 arrives at the desired position at about 6 s, the curves in Figures 7.5 and 7.7 illustrate that the two angular deviations are not eradicated until 10 s. In order to keep the two angular deviations as small as possible, the driven force f in Figure 7.9 sharply changes at the outset. When the trolley is close to its desired position, f is mildly applied to the double-pendulum-type system. As displayed in Figures 7.5–7.8, the hook and payload angles still have some tiny residual oscillations when the trolley arrives at the desired position.

Compared with the results in Chapter 3, results in Figures 7.5–7.8 indicate that the residual oscillations by the inside-the-loop control design are smaller than the sole passivity-based anti-sway control. Furthermore, the result in Figure 7.9 illustrates that

Figure 7.3: Trolley position versus time by the integration of the convolved input-shaping-based control and the passivity-based control.

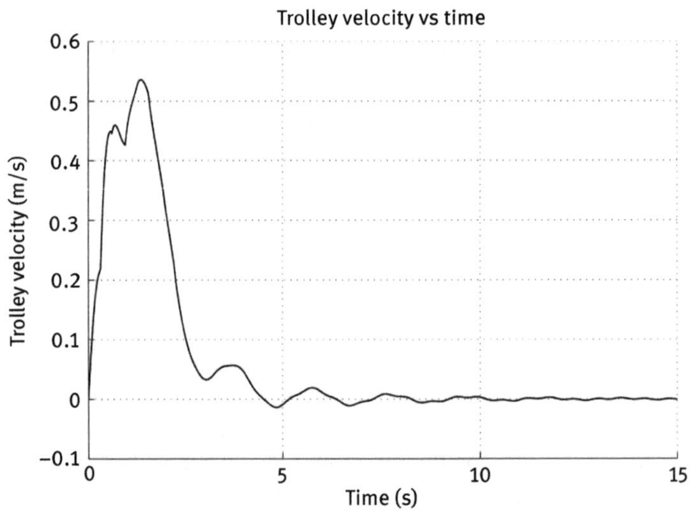

Figure 7.4: Trolley velocity versus time by the integration of the convolved input-shaping-based control and the passivity-based control.

the shaper in the hybrid control design acts as a limiter as well. When the double-pendulum crane moves, the shaper not only shapes the control signal via the proposed control design but also it can decrease the fluctuation of the control signal applied to the trolley.

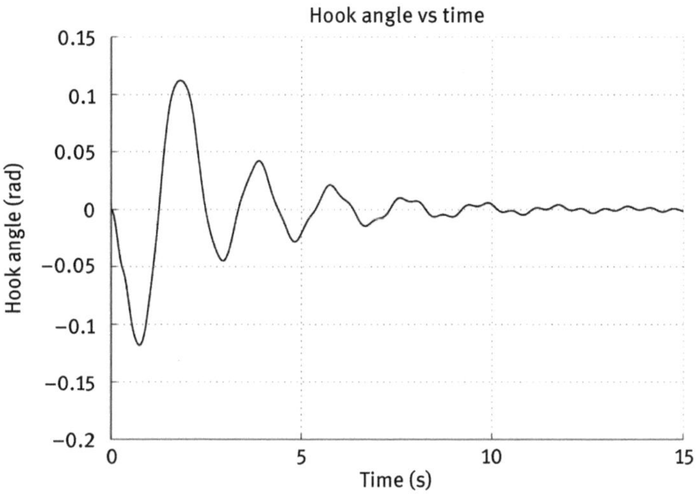

Figure 7.5: Hook angle versus time by the integration of the convolved input-shaping-based control and the passivity-based control.

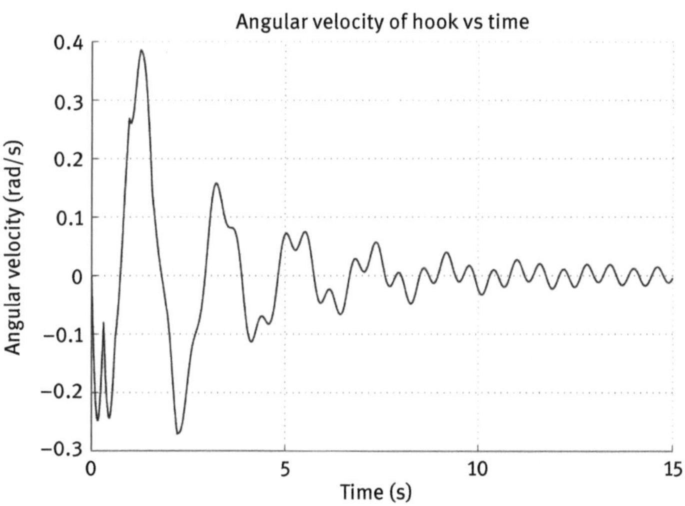

Figure 7.6: Angular velocity of hook versus time by the integration of the convolved input-shaping-based control and the passivity-based control.

7.3.2 Simultaneous Input-Shaping-Based Anti-sway Control

Adapt the structure in Figure 7.2 and integrate the simultaneous input-shaping-based anti-sway control and the passivity-based control. Such a control design is presented to realize the transport control of double-pendulum-type overhead crane systems.

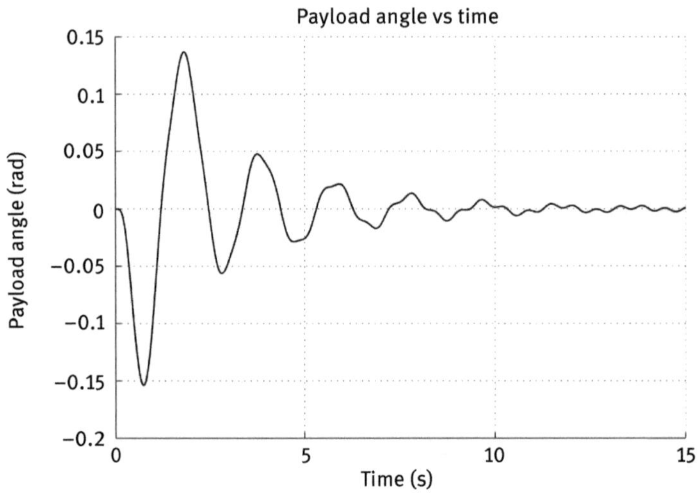

Figure 7.7: Payload angle versus time by the integration of the convolved input-shaping-based control and the passivity-based control.

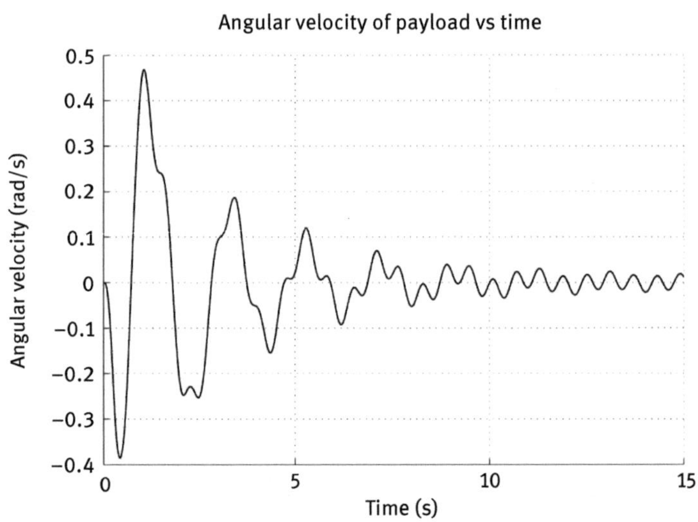

Figure 7.8: Angular velocity of payload versus time by the integration of the convolved input-shaping-based control and the passivity-based control.

Without loss of generality, the physical parameters of the double-pendulum-type crane system are kept uncharged, that is, $m = 0.5$ kg, $m_1 = 0.25$ kg, $m_2 = 0.5$ kg, $l_1 = 0.5$ m, and $l_2 = 0.5$ m.

The passivity-based controller [9] proposed in Chapter 3 is employed here. The simultaneous input shaper has been designed in eq. (7.15). Finally, the control law applied to the crane system is determined by eq. (7.16). Note that one value in eq. (7.15)

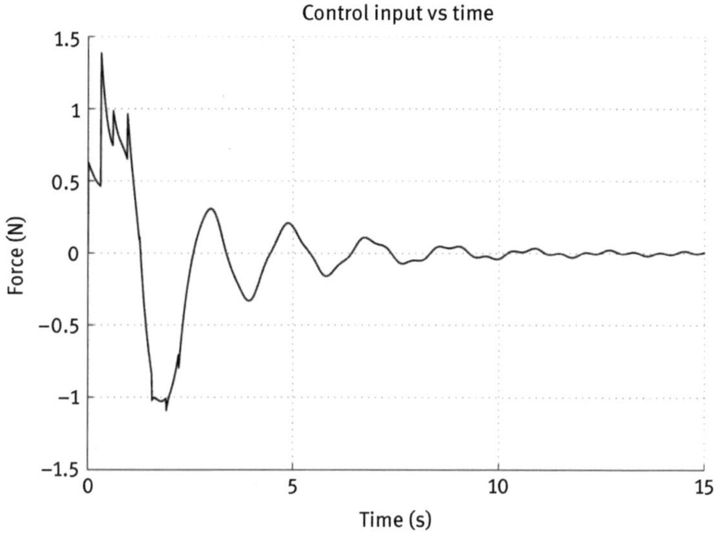

Figure 7.9: Control input versus time by the integration of the convolved input-shaping-based control and the passivity-based control.

is zero, which means the impulse is kept uncharged at that time. Therefore, the series of impulses applied to the crane system is

$$\mathrm{SIS}_{\mathrm{shaper}} = \begin{bmatrix} A_i \\ t_i \end{bmatrix} = \begin{bmatrix} 0.2152 & 0.2848 & 0.2848 & 0.2152 \\ 0 & 0.2382 & 0.8727 & 1.2290 \end{bmatrix}$$

The series of impulses can also guarantee that the crane system has the minimum t_n and the false time is abandoned.

The control objective is to transport the payload to a desired position. The desired position x_d is located at 1 m, that is, $x_d = 1$ m from the origin in Figure 2.2. Meanwhile, the controller needs to suppress the double-pendulum motions of the hook and payload. The results are presented in Figures 7.10–7.16.

As displayed in Figures 7.10–7.16, the inside-the-loop design that integrates the controller and the shaper is able to achieve the transport control objective. The curve in Figure 7.10 arrives at the desired position at about 6 s. But the curves in Figures 7.12 and 7.14 illustrate that the two angular deviations still have some residual oscillations. From the simulation results, the simultaneous input shaper does not have a better performance than the convolved input shaper. In practice, it is very hard to accurately model the frequencies of double-pendulum motions. The sight frequency errors may dramatically decrease the control performance. In contrast with the convolved input shaper, the simultaneous input shaper has a constraint about the residual oscillations. The constraint can endure some frequency errors. In real applications, the simultaneous input shaper often has more robust performance than the convolved input shaper.

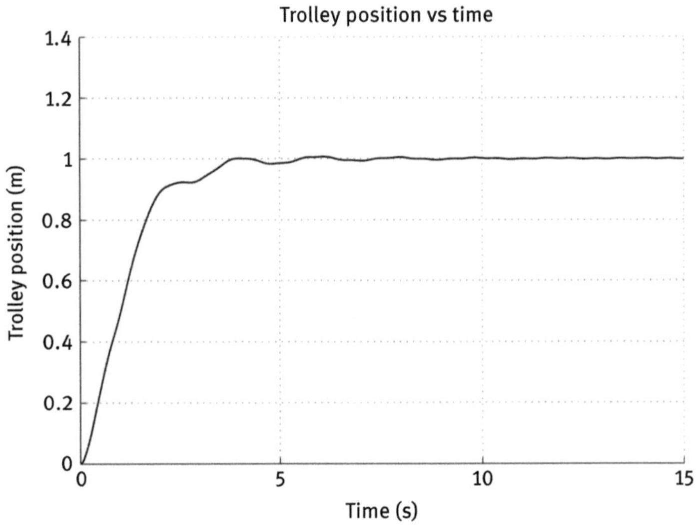

Figure 7.10: Trolley position versus time by the integration of the simultaneous input-shaping-based control and the passivity-based control.

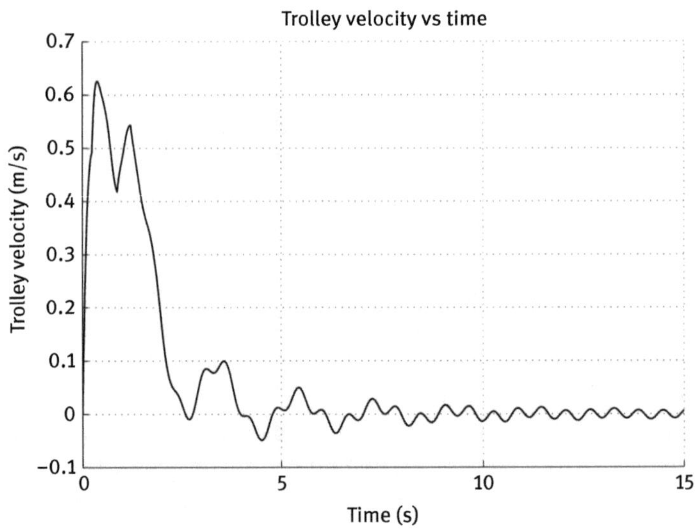

Figure 7.11: Trolley velocity versus time by the integration of the simultaneous input-shaping-based control and the passivity-based control.

7.3.3 Input-Shaping-Based Fuzzy Anti-sway Control

In order to improve the performance of the simultaneous input shaper, the structure in Figure 7.2 is adapted and the simultaneous input-shaping-based anti-sway control

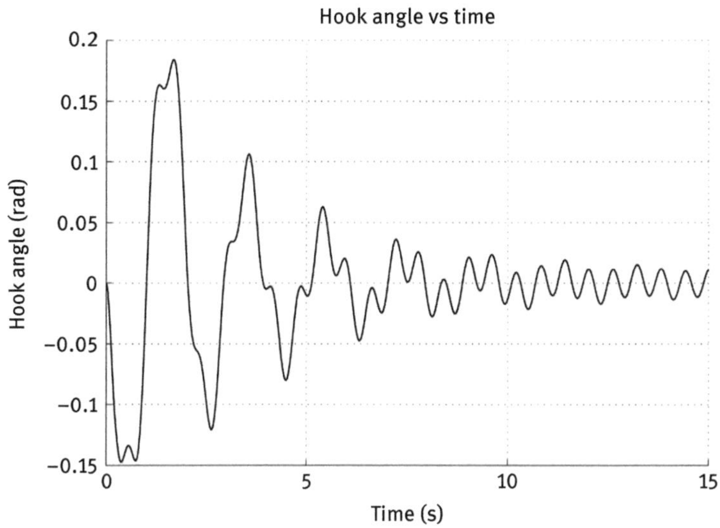

Figure 7.12: Hook angle versus time by the integration of the simultaneous input-shaping-based control and the passivity-based control.

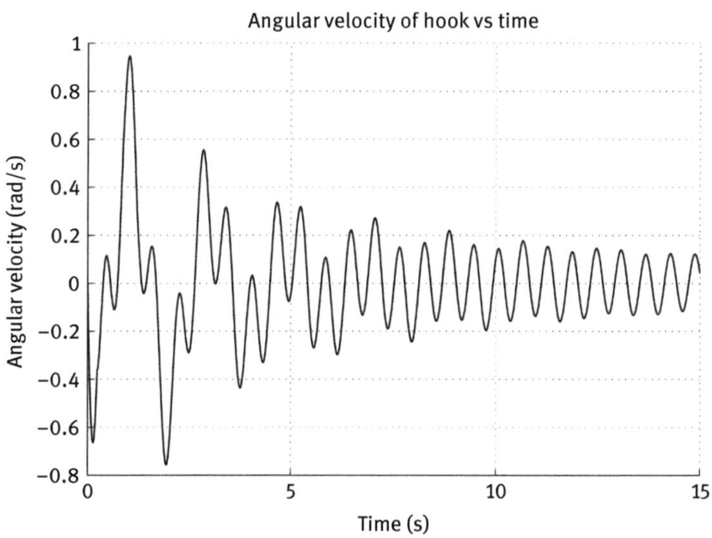

Figure 7.13: Angular velocity of hook versus time by the integration of the simultaneous input-shaping-based control and the passivity-based control.

is integrated with the passivity-based control. Such a control design is presented to realize the transport control of double-pendulum-type overhead crane systems.

Without loss of generality, physical parameters of the double-pendulum-type crane system are kept uncharged, that is, $m = 0.5$ kg, $m_1 = 0.25$ kg, $m_2 = 0.5$ kg,

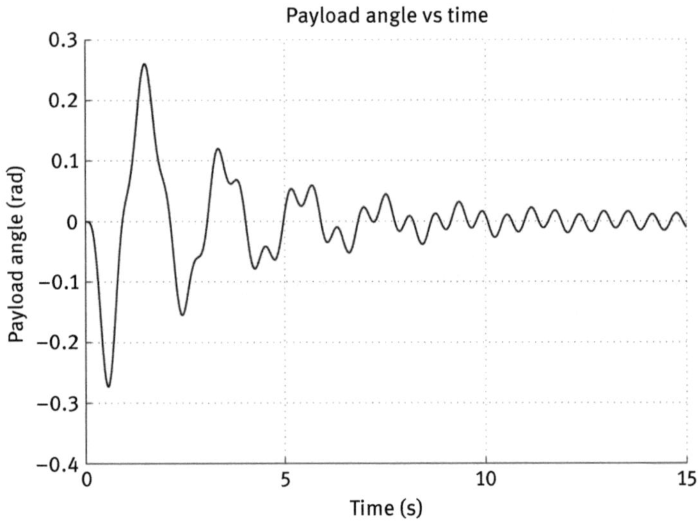

Figure 7.14: Payload angle versus time by the integration of the simultaneous input-shaping-based control and the passivity-based control.

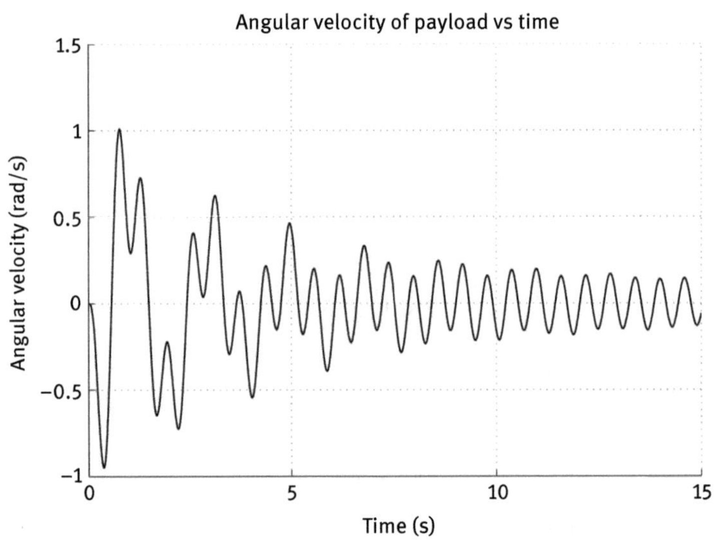

Figure 7.15: Angular velocity of payload versus time by the integration of the simultaneous input-shaping-based control and the passivity-based control.

$l_1 = 0.5$ m, and $l_2 = 0.5$ m. The single-input-rule module (SIRM)-based fuzzy controller proposed in Chapter 6 is employed here. The simultaneous input shaper has been designed in eq. (7.15). Finally, the control law applied to the crane system is determined by eq. (7.16), where the false impulse is abandoned as well.

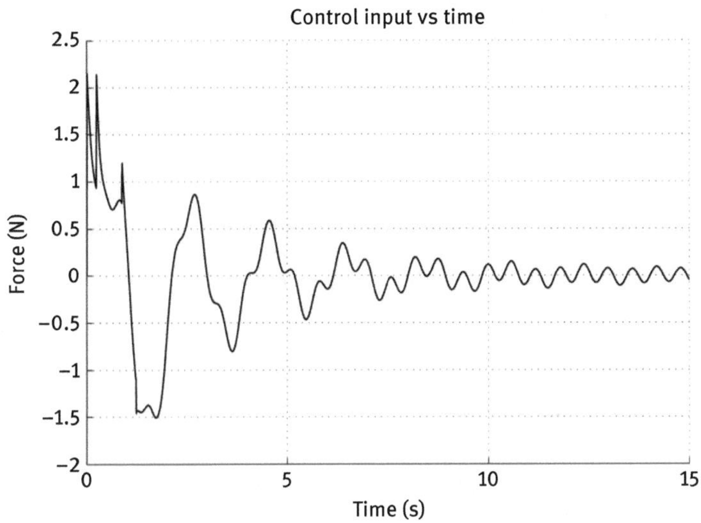

Figure 7.16: Control input versus time by the integration of the simultaneous input-shaping-based control and the passivity-based control.

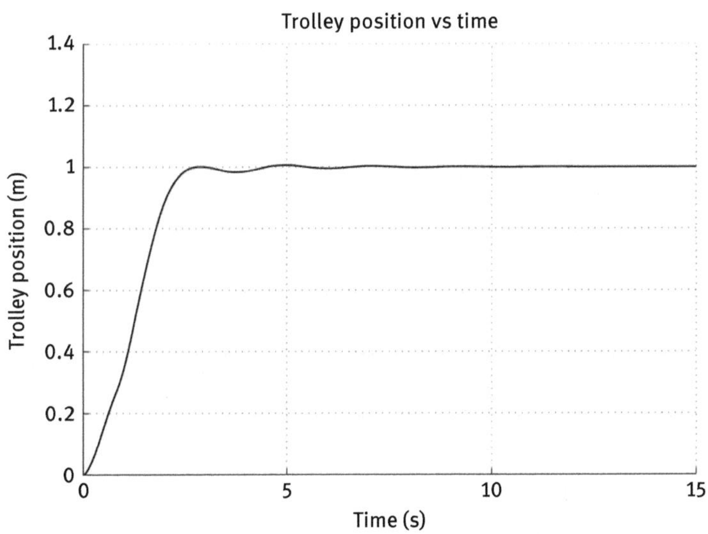

Figure 7.17: Trolley position versus time by the integration of the simultaneous input-shaping-based control and the SIRM-based fuzzy control.

The control objective is to transport the payload to a desired position. The desired position x_d is located at 1 m, that is, $x_d = 1$ m from the origin in Figure 2.2. Meanwhile, the controller needs to suppress the double-pendulum motions of the hook and payload. The results are presented in Figures 7.17–7.23.

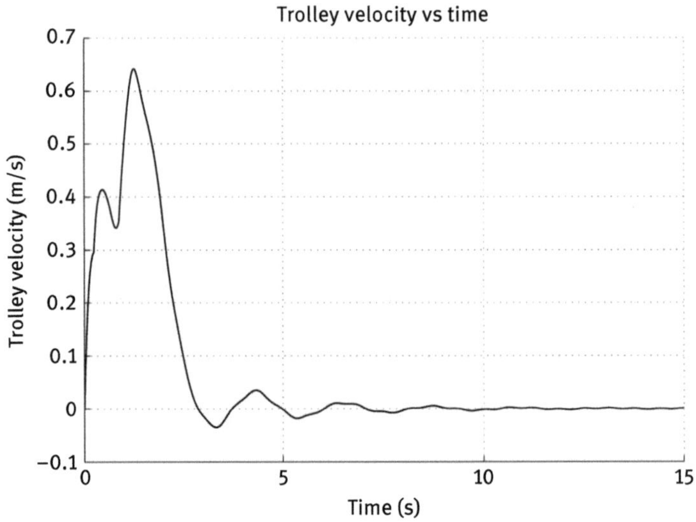

Figure 7.18: Trolley velocity versus time by the integration of the simultaneous input-shaping-based control and the SIRM-based fuzzy control.

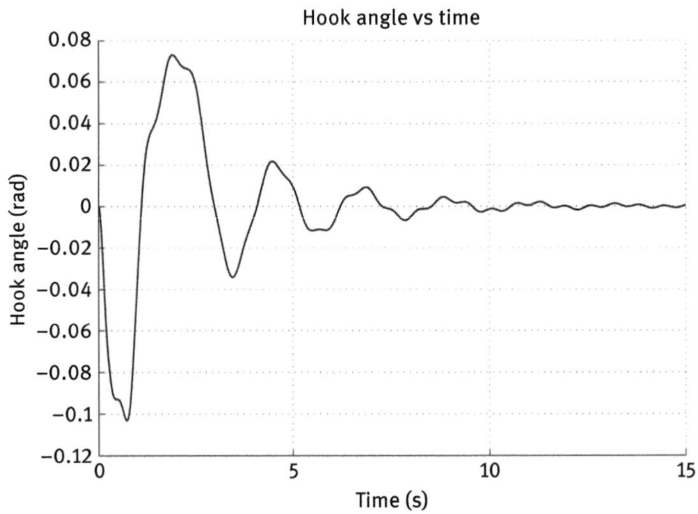

Figure 7.19: Hook angle versus time by the integration of the simultaneous input-shaping-based control and the SIRM-based fuzzy control.

As displayed in Figures 7.17–7.23, the inside-the-loop design that integrates the fuzzy controller and the shaper is able to achieve the transport control objective. The curve in Figure 7.17 arrives at the desired position at about 5 s. Compared with the hook-angle and payload-angle curves in Figures 7.12 and 7.14, Figures 7.19 and 7.21 illustrate that the two angular deviations still have smaller residual oscillations. Meanwhile,

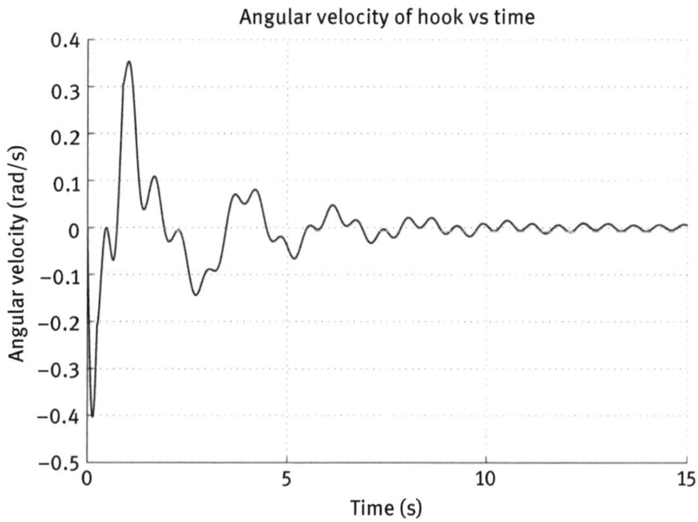

Figure 7.20: Angular velocity of hook versus time by the integration of the simultaneous input-shaping-based control and the SIRM-based fuzzy control.

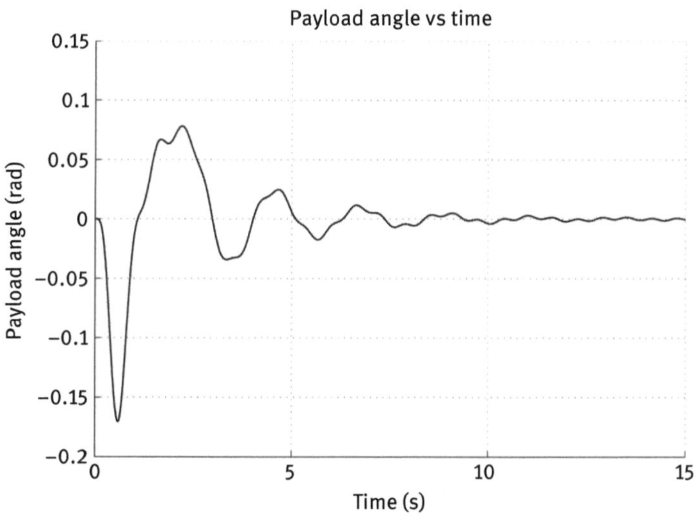

Figure 7.21: Payload angle versus time by the integration of the simultaneous input-shaping-based control and the SIRM-based fuzzy control.

the control input in Figure 7.23 is also improved. From the simulation results, the integration of the simultaneous input shaper and SIRM-based fuzzy controller has a better performance than the integration of the passivity-based controller and the simultaneous input shaper. The fact means the integration of the fuzzy controller and the shaper contributes to the improvement of the control performance.

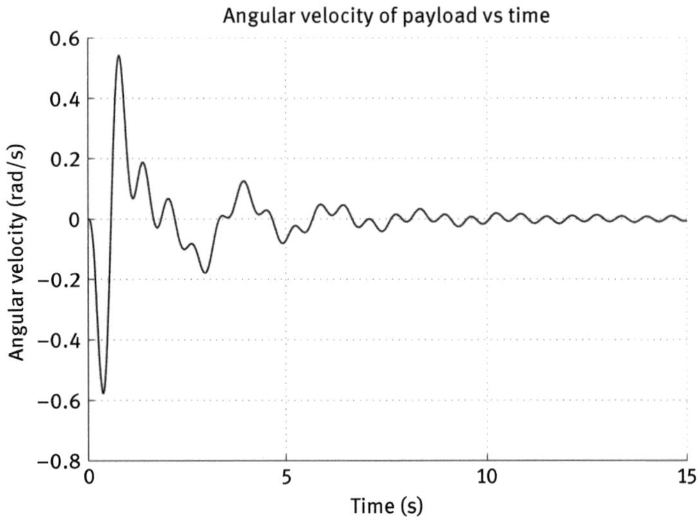

Figure 7.22: Angular velocity of payload versus time by the integration of the simultaneous input-shaping-based control and the SIRM-based fuzzy control.

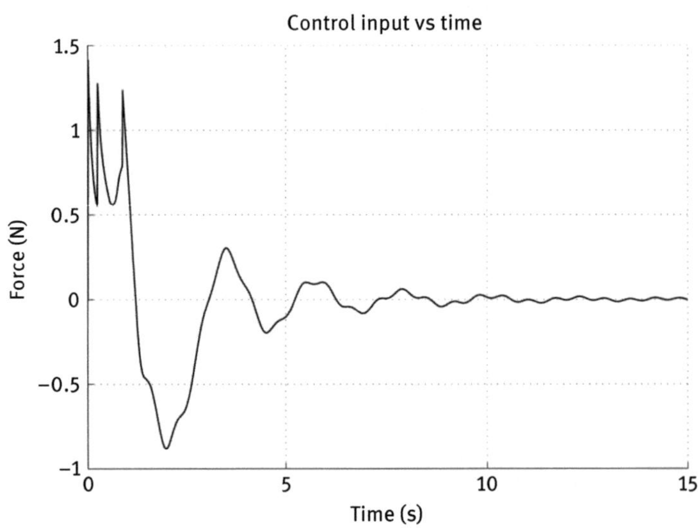

Figure 7.23: Control input versus time by the integration of the simultaneous input-shaping-based control and the SIRM-based fuzzy control.

On point should be claimed that the fuzzy controller needs six state variables, that is x, θ_1, θ_2 and their derivatives, to determine the final control signal. In contrast, the passivity-based controller needs only two state variables, that is, x and its derivative, to calculate the final control signal. In practice, these state variables have to be measured by sensors. How and where to arrange the sensors is subjected by the crane structure. From the perspective, the integration of the passivity-based controller and the shaper is more user-friendly because it is very easy to measure x and its derivative in reality. However, the measurement of θ_1, θ_2 and their derivatives is rather difficult because it is very hard to arrange the angular sensors on the double-pendulum-type overhead crane subjected by the mechanical structure.

7.4 Source Codes

The source codes [10] to draw the simultaneous input shaper are listed as follows, as well as the codes to plot Figures 7.3–7.23:

Matlab codes to solve the optimization problem for the simultaneous input shaper.

Main function entitled 'SISshaper.m'

```
clc;
%x0=[0.2151;0.2849;-0.0000;0.2849;0.2151;0.2381;0.8833;
           0.9909;1.2290]; % the final result
%x0=[0.2151;0.2849;-0.0000;0.2849;0.2151;0.2381;
           0.8833;0.9909;1.2290]; % the result of the
           third iteration
%x0=[0.0539;0.4198;0.0526;0.4198;0.0539;-0;0.8322;
           0.9322;1.3949;];
           % the result of the second iteration
%x0=[0.2152;0.2848;0;0.2848;0.2152;
           0.2382;0.8727;0.9908;1.229]; % the result of the
           first iteration
x0=[0.1;0.2;0.4;0.2;0.1;0.63;1.26;1.89;2.52];
           % Initial value
Aeq=[1 1 1 1 1 0 0 0 0;1 0 0 0 -1 0 0 0 0;0 1 0
    -1 0 0 0 0 0];
beq=[1;0;0];
VLB=[0;0;0;0;0;0;0;0;0];
VUB=[1;1;1;1;1;3;3;3;3];
```

```
A=[-1 0 0 0 0 0 0 0 0;
    0 -1 0 0 0 0 0 0 0;
    0 0 -1 0 0 0 0 0 0;
    0 0 0 -1 0 0 0 0 0;
    0 0 0 0 -1 0 0 0 0;
    0 0 0 0 0 -1 0 0 0;
    0 0 0 0 0 1 -1 0 0;
    0 0 0 0 0 0 1 -1 0;
    0 0 0 0 0 0 0 1 -1];
b=[0;0;0;0;0;0.1;-0.1; -0.1; -0.1];

options=optimset('LargeScale','off','display','on');
[x,fvel]=fmincon(@objfun,x0,A,b,Aeq,beq,VLB,VUB,@confun,
         options);
```

Objective function entitled 'objfun.m'

```
function f=objfun(x)
f=x(9);
```

Constraints function entitled 'confun.m'

```
function [c, ceq]=confun(x)
g=9.81;
l1=0.5;
l2=0.5;
R=2; %R=m2/m1

beta=sqrt((1+R)^2*(1/l1+1/l2)^2-4*((1+R)/l1/l2));

w1=sqrt(g/2)*sqrt((1+R)*(1/l1+1/l2)-beta);
w2=sqrt(g/2)*sqrt((1+R)*(1/l1+1/l2)+beta);

alpha=-g*(1+R)/(w1*w2*l1*l2);
k0=beta*l1*g;
```

```
k1=w1*l1*(w2^2*alpha*(l1+l2))/k0;
k2=-w2*l1*(w1^2*alpha*(l1+l2))/k0;

k=abs(k1*sqrt((x(1)*cos(w1*0)+x(2)*cos(w1*x(6))+x(3)*
  cos(w1*x(7))+x(4)*cos(w1*x(8))+x(5)*cos(w1*x(9)))^2
  +(x(1)*sin(w1*0)+x(2)*sin(w1*x(6))+x(3)*sin(w1*x(7))
  +x(4)*sin(w1*x(8))+x(5)*sin(w1*x(9)))^2))+abs(k2*
  sqrt((x(1)*cos(w2*0)+x(2)*cos(w2*x(6))+x(3)*cos(w2*x(7))
  +x(4)*cos(w2*x(8))+x(5)*cos(w2*x(9)))^2
  +(x(1)*sin(w2*0)+x(2)*sin(w2*x(6))+x(3)*sin(w2*x(7))
  +x(4)*sin(w2*x(8))+x(5)*sin(w2*x(9)))^2));
c=k-0.05;
ceq=[];
```

Matlab codes to plot Figures 7.3–7.9

```
function Inputshaping
clear all;
close all;
len1=0.5; %1/2physical parameters
len2=0.5;
mass=0.5;
mass1=0.25;
mass2=0.5;
g=9.81;

s=0; %initial conditions
v=0;
th1=0;
th2=0;
w1=0;
w2=0;
f=0;
t=0;
```

```
s_1=s;
v_1=v;
th_1=th1;
th_2=th2;
w_1=w1;
w_2=w2;
dt=0.01; % sampling time
xend=1; % desired position

sdata=s;
vdata=v;
th1data=th1;
th2data=th2;
w1data=w1;
w2data=w2;
fdata=f;
tdata=t;

for i=1:1:1500
s=s_1;
v=v_1;
th1=th_1;
th2=th_2;
w1=w_1;
w2=w_2;
M=[mass+mass1+mass2 (mass1+mass2)*len1 mass2*len2;...
      % system model
  (mass1+mass2)*len1 (mass1+mass2)*len1^2 mass2*
      len1*len2;...mass2*len2 mass2*len1*len2 mass2*len2^2];

C=[0 -1*(mass1+mass2)*len1*w1*th1 -1*mass2*len2*w2*th2;...
   0 0 mass2*len1*len2*w1*(th1-th2);...
   0 -1*mass2*len1*len2*w1*(th1-th2) 0];
G=[0;(mass1+mass2)*g*len1*th1;mass2*g*len2*th2];
```

```
f=controller(xend,s,v,len1,len2,mass1,mass2,g,t);
T=[f;0;0];
x=inv(M)*(T-G-C*[v;w1;w2]);
y=[v;w1;w2]+dt*x;
v_1=y(1);
w_1=y(2);
w_2=y(3);
z=[s;th1;th2]+0.5*dt^2*x+dt*[v;w1;w2];
s_1=z(1);
th_1=z(2);
th_2=z(3);
t=t+dt;
tdata = [tdata,t];
fdata = [fdata,f];
vdata = [vdata,v_1];
th1data=[th1data,th_1];
th2data=[th2data,th_2];
w1data=[w1data,w_1];
w2data=[w2data,w_2];
sdata=[sdata,s_1];
end

h1=figure(1);
set(h1,'Position',[100,200,500,300],...
      'Menu','none',...
      'NumberTitle','off',...
      'Name','Trolley position'...
);
axes1=axes('Parent',h1);
title(axes1,'Trolley position vs time','FontSize',16);
xlabel(axes1,'Time(s)');
ylabel(axes1,'Trolley position(m)');
hold(axes1,'all');
plot(tdata,sdata);
grid on;
```

```
h1=figure(2);
set(h1,'Position',[100,200,500,300],...
     'Menu','none',...
     'NumberTitle','off',...
     'Name','Trolley velocity'...
);
axes1=axes('Parent',h1);
title(axes1,'Trolley velocity vs time',
    'FontSize',16);
xlabel(axes1,'Time(s)');
ylabel(axes1,'Trolley velocity(m/s)');
hold(axes1,'all');
plot(tdata,vdata);
grid on;

h2=figure(3);
set(h2,'Position',[800,200,500,300],...
     'Menu','none',...
     'NumberTitle','off',...
     'Name','Hook angle'...
);
axes2=axes('Parent',h2);
title(axes2,'Hook angle vs time','FontSize',16);
xlabel(axes2,'Time(s)');
ylabel(axes2,'Hook angle(rad)');
hold(axes2,'all');
plot(tdata,th1data,'r');
grid on;

h2=figure(4);
set(h2,'Position',[800,200,500,300],...
     'Menu','none',...
     'NumberTitle','off',...
     'Name','Angular velocity'...
);
```

```
axes2=axes('Parent',h2);
title(axes2,'Angular velocity of hook vs time',
    'FontSize',16);
xlabel(axes2,'Time(s)');
ylabel(axes2,'Angular velocity(rad/s)');
hold(axes2,'all');
plot(tdata,w1data,'r');
grid on;

h2=figure(5);
set(h2,'Position',[800,200,500,300],...
     'Menu','none',...
     'NumberTitle','off',...
     'Name','Payload angle'...
);
axes2=axes('Parent',h2);
title(axes2,'Payload angle vs time','FontSize',16);
xlabel(axes2,'Time(s)');
ylabel(axes2,'Payload angle(rad)');
hold(axes2,'all');
plot(tdata,th2data,'r');
grid on;

h2=figure(6);
set(h2,'Position',[800,200,500,300],...
     'Menu','none',...
     'NumberTitle','off',...
     'Name','Angular velocity'...
);
axes2=axes('Parent',h2);
title(axes2,'Angular velocity of payload vs time',
     'FontSize',16);
xlabel(axes2,'Time(s)');
ylabel(axes2,'Angular velocity(rad/s)');
```

```
hold(axes2,'all');
plot(tdata,w2data,'r');
grid on;

h2=figure(7);
set(h2,'Position',[800,200,500,300],...
     'Menu','none',...
     'NumberTitle','off',...
     'Name','Control input'...
);
axes2=axes('Parent',h2);
title(axes2,'Control input vs time','FontSize',16);
xlabel(axes2,'Time(s)');
ylabel(axes2,'Force(N)');
hold(axes2,'all');
plot(tdata,fdata,'r');
grid on;

function f = controller(xend,s,v,len1,len2,mass1,mass2,g,t)
  k1=10; % passivity-base controller
  k2=10;

  l1=0.5;
  l2=0.5;
  R=2; %m2/m1

  beta=sqrt((1+R)^2*(1/l1+1/l2)^2-4*((1+R)/l1/l2));

  fw2=sqrt(g/2)*sqrt((1+R)*(1/l1+1/l2)-beta); %angular
      frequencies
  fw1=sqrt(g/2)*sqrt((1+R)*(1/l1+1/l2)+beta);

  ro=pi/fw1;
  ro1=pi/fw2;
```

```
if t<ro % the convolved input shaper and the
     passivity-based controller
        f=1/16*(k1*(xend-s)-k2*v);
     elseif t>=ro&&t<2*ro
        f=(1/16+1/8)*(k1*(xend-s)-k2*v);
     elseif t>=2*ro&&t<ro1
        f=(1/16+1/8+1/16)*(k1*(xend-s)-k2*v);
     elseif t>=ro1&&t<ro+ro1
        f=(1/16+1/8+1/16+1/8)*(k1*(xend-s)-k2*v);
     elseif t>=ro+ro1&&t<2*ro+ro1
        f=(1/16+1/8+1/16+1/8+1/4)*(k1*(xend-s)-k2*v);
     elseif t>=2*ro+ro1&&t<2*ro1
        f=(1/16+1/8+1/16+1/8+1/4+1/8)*(k1*(xend-s)-k2*v);
     elseif t>=2*ro1&&t<2*ro1+ro
        f=(1/16+1/8+1/16+1/8+1/4+1/8+1/16)*
          (k1*(xend-s)-k2*v);
     elseif t>=2*ro1+ro&&t<2*ro1+2*ro
        f=(1/16+1/8+1/16+1/8+1/4+1/8+1/16+1/8)*
          (k1*(xend-s)-k2*v);
     elseif t>=2*ro1+2*ro
        f=(1/16+1/8+1/16+1/8+1/4+1/8+1/16+1/8+1/16)*
          (k1*(xend-s)-k2*v);
  end
```

Matlab codes of the control design to plot Figures 7.10–7.16

```
function f = controller(xend,s,v,len1,len2,mass1,mass2,g,t)
  k1=10;
  k2=10;

  A1=0.2152;A2=0.2848;A3=0.2848;A4=0.2152;
  t1=0.2382;t2=0.8727;t3=1.2290;
  if t<t1
      f=(A1)*(k1*(xend-s)-k2*v);
      elseif t>=t1&&t<t2
```

```
            f=(A1+A2)*(k1*(xend-s)-k2*v);
             elseif t>t2&&t<t3
               f=(A1+A2+A3)*(k1*(xend-s)-k2*v);
                elseif t>=t3
                   f=(A1+A2+A3+A4)*(k1*(xend-s)-k2*v);
    end
```

Matlab codes of the control design to plot Figures 7.17–7.23

```
function f = controller(mass,mass1,mass2,xend,s,v,th1,
   th2,w1,w2,t )

a1=th1/0.5;
b1=w1/1.5;
if a1>-1&&a1<0
   u1=(a1^2+2*a1)/(-2*a1^2-2*a1+1);
elseif a1>0&&a1<1
   u1=(2*a1-a1^2)/(-2*a1^2+2*a1+1);
elseif a1<-1
   u1=-1;
elseif a1==0
   u1=0;
elseif a1>1
   u1=1;
end
if b1>-1&&b1<0
   u2=(b1^2+2*b1)/(-2*b1^2-2*b1+1);
elseif b1>0&&b1<1
   u2=(2*b1-b1^2)/(-2*b1^2+2*b1+1);
elseif b1<-1
   u2=-1;
elseif b1==0
   u2=0;
elseif b1>1
```

```
    u2=1;
end
a2=th2/0.5;
b2=w2/1.5;
if a2>-1&&a2<0
    u3=(a2^2+2*a2)/(-2*a2^2-2*a2+1);
elseif a2>0&&a2<1
    u3=(2*a2-a2^2)/(-2*a2^2+2*a2+1);
elseif a2<-1
    u3=-1;
elseif a2==0
    u3=0;
elseif a2>1
    u3=1;
end
if b2>-1&&b2<0
    u4=(b2^2+2*b2)/(-2*b2^2-2*b2+1);
elseif b2>0&&b2<1
    u4=(2*b2-b2^2)/(-2*b2^2+2*b2+1);
elseif b2<-1
    u4=-1;
elseif b2==0
    u4=0;
elseif b2>1
    u4=1;
end
if (xend-s)>=-1&&(xend-s)<0
    u5=((xend-s)^2+2*(xend-s))/(-2*(xend-s)^2-2*(xend-s)+1);
elseif (xend-s)>0&&(xend-s)<=1
    u5=(2*(xend-s)-(xend-s)^2)/(-2*(xend-s)^2+2*(xend-s)+1);
elseif (xend-s)<-1
    u5=-1;
elseif (xend-s)==0
    u5=0;
```

```
elseif (xend-s)>1
   u5=1;
end
if v>-1&&v<0
   u6=(-1*v^2-2*v)/(-2*v^2-2*v+1);
elseif v>0&&v<1
   u6=(v^2-2*v)/(-2*v^2+2*v+1);
elseif v<-1
   u6=1;
elseif v==0
   u6=0;
elseif v>1
   u6=-1;
end
if th1>-0.5&&th1<0.5
   r1=(abs(th1)-abs(th1)^2)/(0.5+2*abs(th1)-4*abs(th1)^2);
   r2=1;
elseif abs(th1)>=0.5&&abs(th1)<=1
   r1=(5*abs(th1)-3*abs(th1)^2-1.5)/(6*abs(th1)
      -4*abs(th1)^2-1.5);
   r2=0;
elseif abs(th1)>1
   r1=1;
   r2=0;
end

   a3=max(abs(a1),abs(a2));
   b3=min(abs(a1),abs(a2));

if abs(a1)>=0.5&&abs(a1)<=1&&abs(a2)>=0.5&&abs(a2)<=1
   r3=0;
elseif a1>-0.5&&a1<0.5&&a2>-0.5&&a2<0.5
   r3=(a3-a3^2+0.5-2*b3^2)/(2*a3-2*a3^2+0.5-2*b3^2);
elseif a1>-0.5&&a1<0.5&&abs(a2)>=0.5&&abs(a2)<=1
```

```
    if (0.5-abs(a1))<=(abs(a2)-0.5)
        r3=(abs(a1)-3*a1^2+0.5)/(4*abs(a2)-2*a2^2
            -4*a1^2+2*abs(a1)-1);
    elseif (0.5-abs(a1))>(abs(a2)-0.5)
        r3=(0.5-2*a1^2+abs(a2)-a2^2)/(6*abs(a2)
            -4*a2^2-2*a1^2-1);
    end
elseif a2>-0.5&&a2<0.5&&abs(a1)>=0.5&&abs(a1)<=1
    if (0.5-abs(a2))<=(abs(a1)-0.5)
        r3=(abs(a2)-3*a2^2+0.5)/(4*abs(a1)-2*a1^2
             -4*a2^2+2*abs(a2)-1);
    elseif (0.5-abs(a2))>(abs(a1)-0.5)
        r3=(0.5-2*a2^2+abs(a1)-a1^2)/(6*abs(a1)-4*a1^2
            -2*a2^2-1);
    end
elseif abs(a1)>1||abs(a2)>1
    r3=0;
end

A1=0.2152;A2=0.2848;A3=0.2848;A4=0.2152;
t1=0.2382;t2=0.8727;t3=1.2290;
if t<t1
f=(A1)*(1.5*(mass+mass1+mass2)*((0.7+0.3*r1)*(u1+u2)
  +(0.2*r2)*(u3+u4)+(1.3+2.2*r3)*(u5+u6)));
elseif t>=t1&&t<t2
f=(A1+A2)*(1.5*(mass+mass1+mass2)*((0.7+0.3*r1)*(u1+u2)
  +(0.2*r2)*(u3+u4)+(1.3+2.2*r3)*(u5+u6)));
elseif t>t2&&t<t3
f=(A1+A2+A3)*(1.5*(mass+mass1+mass2)*((0.7+0.3*r1)*(u1+u2)
  +(0.2*r2)*(u3+u4)+(1.3+2.2*r3)*(u5+u6)));
elseif t>=t3
f=(A1+A2+A3+A4)*(1.5*(mass+mass1+mass2)*((0.7+0.3*r1)*(u1+u2)
  +(0.2*r2)*(u3+u4)+(1.3+2.2*r3)*(u5+u6)));
end
```

References

[1] Singhose W. Command shaping for flexible systems: A review of the first 50 years. Int J Precis Eng Manuf, 2009, 10(4): 153–168.

[2] Garrido S, Abderrahim M, Giménez A, Diez R, Balaguer C. Anti-swinging input shaping control of an automatic construction crane. IEEE Trans Autom Sci Eng, 2008, 5(3): 549–557.

[3] Vaughan J, Kim D, Singhose W. Control of tower cranes with double-pendulum payload dynamics. IEEE Trans Control Syst Technol, 2010, 18(6): 1345–1358.

[4] Singhose W, Kim D, Kenison M. Input shaping control of double-pendulum bridge crane oscillations. J Dyn Syst Meas Control-Trans ASME, 2008, 130(3): DOI:10.1115/1.2907363

[5] Fujioka D, Shah M, Singhose W. Robustness analysis of input-shaped model reference control on a double-pendulum crane. In: Proceedings of American Control Conference, Chicago IL, USA, 2015, pp. 2561–2566.

[6] Pai MC. Robust input shaping control for multi-mode flexible structures using neuro-sliding mode output feedback control. J Frankl Inst – Eng Appl Math, 2012, 349(3): 1283–1303.

[7] Qian DW, Tong SW, Yi JQ. Adaptive control based on incremental hierarchical sliding mode for overhead crane systems. Appl Math Inform Sci, 2013, 7(4): 1359–1364.

[8] Qian D, Tong S, Yang B, Lee S. Design of simultaneous input-shaping-based SIRMs fuzzy control for double-pendulum-type overhead cranes. Bull Pol Acad Sci-Tech Sci, 2015, 63(4): 887–896.

[9] Liu DT, Guo WP, Yi JQ. Dynamics and GA-based stable control for a class of underactuated mechanical systems. Int J Control Autom Syst, 2008, 6(1): 35–43.

[10] Xue D, Chen Y, Atherton DP. Linear feedback control: Analysis and design with MATLAB. Philadelphia, PA: Society for Industrial and Applied Mathematics, 2007.

8 Conclusions

Abstract: This chapter presents preliminary results on the anti-sway control development and application for double-pendulum-type overhead cranes, draws some concluding remarks, and summarizes open problems for future research.

Keywords: Crane control, double-pendulum-type motions, open problems

Overhead crane systems are underactuated because they have a lower number of control inputs than the number of degrees of freedom to be controlled. Under some operating conditions, the hook and payload should be modeled by two distributed mass points rather than a lumped mass point. This distributed-mass-point assumption results in double-pendulum dynamics. The double-pendulum dynamics compound two kinds of pendulum motions with different natural frequencies so that those controllers designed to resist the single-pendulum oscillation are not as effective as expected.

Some control designs and applications have been reported to suppress the double-pendulum oscillations. This book investigates some latest results in this field. Numerical simulations show these popular control methods to resist the challenging double-pendulum motions. In order to feasilize readers, all the source codes using Matlab have been presented and a software program that integrates all the methods of the book has also been provided in the appendix.

The motivation of the book is to investigate the control design for the anti-sway control of double-pendulum-type overhead crane. The technical contents involved in this book focus on control design and control methods. Some specific control problems such as modeling uncertainties and controller tuning are not mentioned. In summary, the control methods of double-pendulum-type overhead cranes can be roughly categorized by three kinds, as follows:

- *Open-loop control.* The representative approaches are the input-shaping-based control techniques. The technical contents have been extensively touched by Prof. W Singhose and his team. Many papers from his group have reported the results. Therefore, the technical contents are not refereed in this book.
- *Closed-loop control.* The representative approaches are the passivity-based control method, the sliding-mode-based control methods, the output feedback-based control, as well as the fuzzy-logic-based control design. The corresponding technical contents are referred in Chapters 3–6, respectively.
- *Compound control.* The compound control methods integrate the open-loop design and the closed-loop methods in order to earn the better control performance. This book has reported the three kinds of the compound control approaches, that is, the integration of the convolved input shaper and the passivity-based

DOI 10.1515/9783110520101-008

controller, the integration of the simultaneous input shaper and the passivity-based controller, and the integration of the simultaneous input shaper and the single-input-rule module-based fuzzy controller. The technical contents have been referred in Chapter 7.

Control of double-pendulum-type overhead crane systems is open and it is difficult to include all the works in this book. However, the author believes that this book can provide readers with the most up-to-date knowledge. Finally, several important issues and possible future research directions are drawn.

- *Solution of the double-pendulum motions of other types of cranes*. There are three types of cranes. This book only takes overhead cranes as an example to illustrate the control design and stability analysis. It is possible and expected to extend these control methods to the other two types of cranes, that is, tower cranes and boom cranes. Since the other two types of cranes are described by non-Cartesian coordinates, some technical details about their control design remain problematic and challenging.
- *Insight into the double-pendulum motions of three-dimensional cranes*. This book presents insight into the double-pendulum motions of a two-dimensional overhead crane, where the linearization technique is adapted to decouple the two dimensions. In a three-dimensional overhead crane, the crane has more complicated coupling in three dimensions and its double-pendulum motions are more challenging and interesting.

Appendix Software Program of Anti-sway Control Against Double-Pendulum Motions

GUI Design

In order to illustrate the control methods, a software program has been developed. The program is based on the graphical user interface (GUI) design of Matlab. The GUI has been developed and demonstrated as follows. Input "double_pendulum" is the command window of Matlab interface. Its source codes are listed as well. The software integrates all the control methods involved in the book.

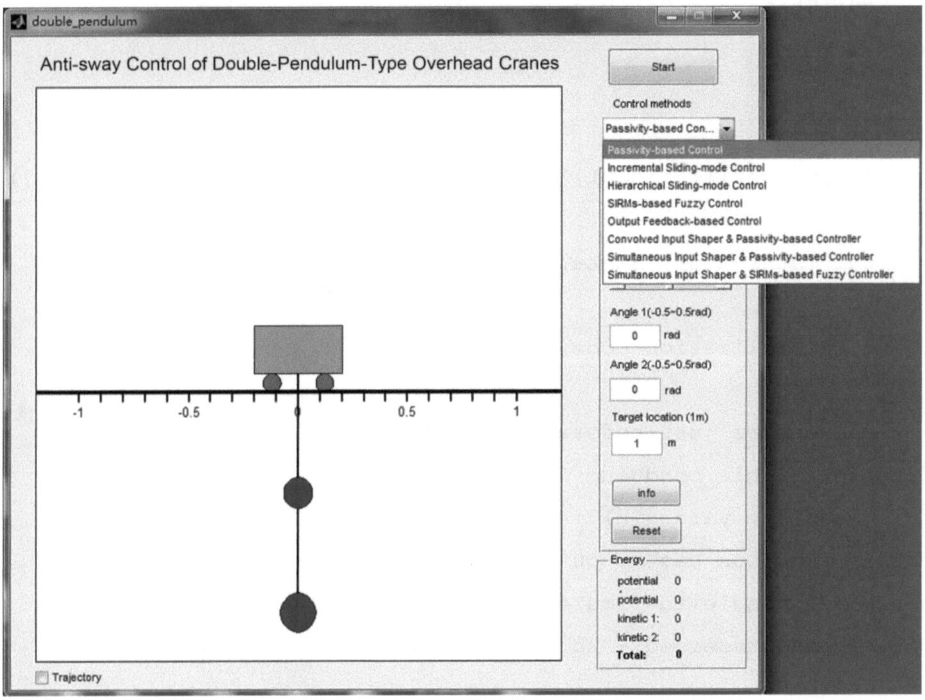

Source Codes of GUI

```
function varargout = double_pendulum(varargin)

% the function simulates double-pendulum type crane systems
%
% Dianwei Qian, Binbin Yang & Qiang Wang, June 2017 (C)
```

DOI 10.1515/9783110520101-009

```
% Begin initialization code - DO NOT EDIT!
gui_Singleton = 1;
gui_State = struct('gui_Name', mfilename, ...
            'gui_Singleton', gui_Singleton, ...
            'gui_OpeningFcn', @double_pendulum_OpeningFcn, ...
            'gui_OutputFcn', @double_pendulum_OutputFcn, ...
            'gui_LayoutFcn', [] , ...
            'gui_Callback', []);
if nargin && ischar(varargin{1})
    gui_State.gui_Callback = str2func(varargin{1});
end

if nargout
    [varargout{1:nargout}] = gui_mainfcn(gui_State, varargin{:});
else
    gui_mainfcn(gui_State, varargin{:});
end
% End initialization code - DO NOT EDIT

% --- Executes just before double_pendulum is made visible.
function double_pendulum_OpeningFcn(hObject, eventdata,
     handles, varargin)
% This function has no output args, see OutputFcn.
% hObject handle to figure
% eventdata reserved - to be defined in a future version of
     MATLAB
% handles structure with handles and user data (see GUIDATA)
% varargin command line arguments to double_pendulum
     (see VARARGIN)

% Choose default command line output for double_pendulum
global u t fdata vdata tdata1 theta1data1 theta2data1 w1data1
       w2data1 sdata1 s1data s2data s3data mode xend f...
    r1 r2 r3 u1 u2 u3 u4 u5 u6 j F F1 s1 s2 s3;
```

```
handles.output = hObject;
handles.len1=.5;
handles.len2=.5;
handles.mass=.5;
handles.mass1=.25;
handles.mass2=.5;
handles.s=0;
handles.v=0;
handles.t1=0;
handles.t2=0;
handles.w1=0;
handles.w2=0;
handles.g=9.8;
xend=1;
u=handles.s;
mode=get(handles.popupmenu1,'Value');
f=0;
r1=0;
r2=0;
r3=0;
u1=0;
u2=0;
u3=0;
u4=0;
u5=0;
u6=0;
t=0;
s1=0;
s2=0;
s3=0;
j=1;
% x_1=0;
F=[0 0 0 0 0 0];
F1=0;
```

```
tdata1=t;
fdata=f;
vdata=handles.v;
theta1data1=handles.t1;
theta2data1=handles.t2;
w1data1=handles.w1;
w2data1=handles.w2;
sdata1=handles.s;
s1data=0;
s2data=0;
s3data=0;
pend_plot(handles);

% Update handles structure
guidata(hObject, handles);

% UIWAIT makes double_pendulum wait for user response
    (see UIRESUME)
% uiwait(handles.figure1);

% --- Outputs from this function are returned to the
        command line.
function varargout = double_pendulum_OutputFcn
        (hObject, eventdata, handles)
% varargout cell array for returning output args (see VARARGOUT);
% hObject handle to figure
% eventdata reserved - to be defined in a future version of
    MATLAB
% handles structure with handles and user data (see GUIDATA)

% Get default command line output from handles structure
varargout{1} = handles.output;

% --- Executes on button press in start_stop.
```

```
function start_stop_Callback(hObject, eventdata, handles)
% hObject handle to start_stop (see GCBO)
% eventdata reserved - to be defined in a future
    version of MATLAB
% handles structure with handles and user data (see GUIDATA)
set(hObject,'string','Stop');
set(handles.theta1,'enable','off');
set(handles.theta2,'enable','off');
set(handles.send,'enable','off');
set(handles.length_rat,'enable','off');
set(handles.mass_rat,'enable','off');
set(handles.reset,'enable','off');
set(handles.gravity,'enable','off');
    global u tdata1 vdata fdata theta1data1 theta2data1
        sdata1 s1data s2data s3data h1 h2 w1data1 w2data1 ueq
        usw mode;
    ueq=0;
    usw=0;
    i=1;
    x_1=0;
    while get(hObject,'value')==1

        if (get(handles.traj,'value'))==1
           x_1=handles.len1*sin(handles.t1);
           y_1=-handles.len1*cos(handles.t1) ;
           _2=handles.len2*sin(handles.t2) + x_1;
           y_2=-handles.len2*cos(handles.t2) + y_1;
           handles.trajectory(1,i)=u+x_2;
           handles.trajectory(2,i)=y_2;
           i=i+1;
           if i>400
              i=1;
           end
  elseif i~=1
```

```
        handles.trajectory=[];
        i=1;
    end

    if mode==5
    [handles.s,handles.v,handles.t1,handles.t2,handles.w1,
           handles.w2,x_1]=...
        Endb_pendulum(handles.t1,handles.t2,handles.w1,
           handles.w2,...
        handles.s,handles.v,handles.g,0.01,x_1);
    else
    [handles.s,handles.v,handles.t1,handles.t2,handles.w1,
           handles.w2]=...
        db_pendulum(handles.t1,handles.t2,handles.w1,
           handles.w2,...
        handles.s,handles.v,handles.mass,handles.mass1,
           handles.mass2,...
        handles.len1,handles.len2,handles.g,.01);
    end
    u=handles.s;
    if u>1
        u=1;
    elseif u<-1
        u=-1;
    end
    pend_plot(handles);
    set(handles.theta1,'string',num2str(handles.t1));
    set(handles.theta2,'string',num2str(handles.t2));
    guidata(hObject, handles);
end

save double_pendulum tdata1 theta1data1 theta2data1
     sdata1 w1data1 w2data1...
vdata fdata s1data s2data s3data
```

```
h1=figure(1); % plot figures
set(h1,'Position',[100,200,500,300],...
      'Menu','none',...
      'NumberTitle','off',...
      'Name','Trolley position'...
);
axes1=axes('Parent',h1);
axis([0 20 -1.2 1.2]);
set(axes1,'xtick',[0 2 4 6 8 10 12 14 16 18 20],'xticklabel',....
      {'0','2','4','6','8','10','12','14','16','18','20'});
set(axes1,'ytick',[-1.2 -1 -0.8 -0.6 -0.4 -0.2 0 0.2
   0.4 0.6 0.8 1 1.2],'yticklabel',....
      {'-1.2','-1','-0.8','-0.6','-0.4','-0.2','0','0.2','0.4',
          '0.6','0.8','1','1.2'});
title(axes1,'Trolley position vs time','FontSize',15);
xlabel(axes1,'Time (s)');
ylabel(axes1,'Trolley position (m)');
hold(axes1,'all');
plot(tdata1,sdata1);
grid on;

h2=figure(2);
set(h2,'Position',[650,100,500,600],...
      'Menu','none',...
      'NumberTitle','off',...
      'Name','Angular outputs'...
);
axes2=subplot(2,1,1);
title(axes2,'Angles vs time','FontSize',15);
axis([0 20 -0.5 0.5]);
set(axes2,'xtick',[0 2 4 6 8 10 12 14 16 18 20],'xticklabel',....
        {'0','2','4','6','8','10','12','14','16','18','20'});
set(axes2,'ytick',[-0.5 -0.4 -0.3 -0.2 -0.1 0 0.1 0.2 0.3 0.4 0.5],
        'yticklabel',....
```

```
        {'-0.5','-0.4','-0.3','-0.2','-0.1','0','0.1','0.2','0.3',
             '0.4','0.5'});
xlabel(axes2,'Time (s)');
ylabel(axes2,'Angle (rad)');
hold(axes2,'all');
plot(tdata1,theta1data1,'r');
hold on;
plot(tdata1,theta2data1,'k');
legend('Hook','Payload');
grid on;
hold on;
axes3=subplot(2,1,2);
title(axes3,'Phase planes of hook and payload','FontSize',15);
xlabel(axes3,'Angle (rad)');
ylabel(axes3,'Angular velocity(rad/s)');
hold(axes3,'all');
plot(theta1data1,w1data1,'r');
hold on;
plot(theta2data1,w2data1,'k');
legend('Hook','Payload');
grid on;

set(hObject,'string','Start');
set(handles.theta1,'enable','on');
set(handles.theta2,'enable','on');
set(handles.send,'enable','on');
set(handles.length_rat,'enable','on');
set(handles.mass_rat,'enable','on');
set(handles.reset,'enable','on');
set(handles.gravity,'enable','on');

% --- Executes on slider movement.
function length_rat_Callback(hObject, eventdata, handles)
% hObject handle to length_rat (see GCBO)
```

```
% eventdata reserved - to be defined in a future
     version of MATLAB
% handles structure with handles and user data (see GUIDATA)
handles.len1=get(hObject,'value');
handles.len2=1-handles.len1;
set(handles.len,'string',num2str(handles.len1));
pend_plot(handles);
guidata(hObject, handles);
% Hints: get(hObject,'Value') returns position of slider
% get(hObject,'Min') and get(hObject,'Max') to determine range
     of slider

% --- Executes during object creation, after setting all
         properties.
function length_rat_CreateFcn(hObject, eventdata, handles)
% hObject handle to length_rat (see GCBO)
% eventdata reserved - to be defined in a future version of
     MATLAB
% handles empty - handles not created until after all
     CreateFcns called

% Hint: slider controls usually have a light gray background.
if isequal(get(hObject,'BackgroundColor'),
   get(0,'defaultUicontrolBackgroundColor'))
   set(hObject,'BackgroundColor',[.9 .9 .9]);
end

% --- Executes on slider movement.
function mass_rat_Callback(hObject, eventdata, handles)
% hObject handle to mass_rat (see GCBO)
% eventdata reserved - to be defined in a future
     version of MATLAB
% handles structure with handles and user data (see GUIDATA)
handles.mass1=get(hObject,'value');
```

```
handles.mass2=1-handles.mass1;
set(handles.mas,'string',num2str(handles.mass1));
pend_plot(handles);
guidata(hObject, handles);
% Hints: get(hObject,'Value') returns position of slider
% get(hObject,'Min') and get(hObject,'Max') to determine range
     of slider

% --- Executes during object creation, after setting all
         properties.
function mass_rat_CreateFcn(hObject, eventdata, handles)
% hObject handle to mass_rat (see GCBO)
% eventdata reserved - to be defined in a future version of
     MATLAB
% handles empty - handles not created until after all
     CreateFcns called

% Hint: slider controls usually have a light gray background.
if isequal(get(hObject,'BackgroundColor'),
   get(0,'defaultUicontrolBackgroundColor'))
   set(hObject,'BackgroundColor',[.9 .9 .9]);
end

function theta1_Callback(hObject, eventdata, handles)
% hObject    handle to theta1 (see GCBO)
% eventdata reserved - to be defined in a future
     version of MATLAB
% handles    structure with handles and user data (see GUIDATA)
handles.t1=str2num(get(hObject,'String'));
pend_plot(handles);
guidata(hObject, handles);
% Hints: get(hObject,'String') returns contents of theta1
     as text
%        str2double(get(hObject,'String')) returns contents of
     theta1 as a double
```

```
% --- Executes during object creation, after setting all
     properties.
function theta1_CreateFcn(hObject, eventdata, handles)
% hObject    handle to text_1 (see GCBO)
% eventdata  reserved - to be defined in a future
     version of MATLAB
% handles    empty - handles not created until after all
     CreateFcns called

% Hint: edit controls usually have a white background
     on Windows.
%       See ISPC and COMPUTER.
if ispc && isequal(get(hObject,'BackgroundColor'),
        get(0,'defaultUicontrolBackgroundColor'))
   set(hObject,'BackgroundColor','white');
end


function theta2_Callback(hObject, eventdata, handles)
% hObject    handle to theta2 (see GCBO)
% eventdata  reserved - to be defined in a future version
     of MATLAB
% handles    structure with handles and user data (see GUIDATA)
handles.t2=str2num(get(hObject,'String'));
pend_plot(handles);
guidata(hObject, handles);
% Hints: get(hObject,'String') returns contents of theta2
     as text
%        str2double(get(hObject,'String')) returns contents
     of theta2 as a double


% --- Executes during object creation, after setting all
         properties.
function theta2_CreateFcn(hObject, eventdata, handles)
```

```
% hObject    handle to theta2 (see GCBO)
% eventdata  reserved - to be defined in a future version of
     MATLAB
% handles    empty - handles not created until after all
     CreateFcns called

% Hint: edit controls usually have a white background on
     Windows.
%       See ISPC and COMPUTER.
if ispc && isequal(get(hObject,'BackgroundColor'),
     get(0,'defaultUicontrolBackgroundColor'))
   set(hObject,'BackgroundColor','white');
end

% --- Executes on button press in traj.
function traj_Callback(hObject, eventdata, handles)
% hObject    handle to traj (see GCBO)
% eventdata  reserved - to be defined in a future version of
     MATLAB
% handles    structure with handles and user data (see GUIDATA)
handles.trajectory=[];
guidata(hObject, handles);
% Hint: get(hObject,'Value') returns toggle state of traj


% --- Executes on button press in reset.
function reset_Callback(hObject, eventdata, handles)
% hObject    handle to reset (see GCBO)
% eventdata  reserved - to be defined in a future version of
     MATLAB
% handles    structure with handles and user data (see GUIDATA)
global u t fdata vdata tdata1 theta1data1 theta2data1 s1data
        s2data s3data sdata1 h1 h2 xend w1data1 w2data1
        mode f...
             r1 r2 r3 u1 u2 u3 u4 u5 u6 j F F1 s1 s2 s3 x_1;
```

```
handles.len1=.5;
handles.len2=.5;
handles.mass=.5;
handles.mass1=.25;
handles.mass2=.5;
handles.t1=0;
handles.t2=0;
handles.w1=0;
handles.w2=0;
handles.s=0;
handles.v=0;
handles.g=9.8;
u=handles.s;
mode=get(handles.popupmenu1,'Value');
f=0;
r1=0;
r2=0;
r3=0;
u1=0;
u2=0;
u3=0;
u4=0;
u5=0;
u6=0;
xend=1;
t=0;
j=1;
x_1=0;
F=[0 0 0 0 0 0];
F1=0;
tdata1=t;
fdata=f;
vdata=handles.v;
```

```
theta1data1=handles.t1;
theta2data1=handles.t2;
w1data1=handles.w1;
w2data1=handles.w2;
sdata1=handles.s;
set(handles.gr,'string','9.8');
set(handles.gravity,'value',9.8);
set(handles.theta1,'string','0');
set(handles.theta2,'string','0');
set(handles.send,'string','1');
set(handles.mas,'string',0.25);
set(handles.mass_rat,'value',0.25);
set(handles.len,'string',0.5);
set(handles.length_rat,'value',0.5);
if ishandle(h1)
close(h1);
end
if ishandle(h2)
close(h2);
end
pend_plot(handles);
guidata(hObject, handles);

function pend_plot(data)
global u;
[xx yy zz]=cylinder(0.1,20);
   x_1=data.len1*sin(data.t1)+u;
   y_1=-data.len1*cos(data.t1) ;
   x_2=data.len2*sin(data.t2) + x_1;
   y_2=-data.len2*cos(data.t2) + y_1;
hold off;
plot([u x_1 x_2],[0 y_1 y_2],'k','linewidth',2);
hold on;
```

```
if get(data.traj,'value') ==1
   plot(data.trajectory(1,:),data.trajectory(2,:),'r.',
        'markersize',3);
end
fill(x_1+xx(1,:).*(data.mass1)^.3,y_1+yy(1,:).*
    (data.mass1)^.3,'b');
fill(x_2+xx(1,:).*(data.mass2)^.3,y_2+yy(1,:).*
    (data.mass2)^.3,'b')
set(gca,'xtick',[],'ytick',[]);
hold on;
patch([-0.2+u -0.2+u 0.2+u 0.2+u],[0 0.2 0.2 0],'g');
hold on;
a = 2*pi*(0:.05:.95)';
xball = 0.04*sin(a)+u;
yball = 0.04*(1+cos(a));
patch(xball-0.12,yball-0.08,'r');
patch(xball+0.12,yball-0.08,'r');
hold on;
plot([-1.2 1.2],[-0.08 -0.08],'LineWidth',3,'Color','k');
i=-1;
while i<=1
plot([i,i],[-0.08,-0.12],'LineWidth',2,'Color','k');
i=i+0.1;
end
text(-1,-0.16,'-1','horizontalalignment','center');
text(-0.5,-0.16,'-0.5','horizontalalignment','center');
text(0,-0.16,'0','horizontalalignment','center');
text(0.5,-0.16,'0.5','horizontalalignment','center');
text(1,-0.16,'1','horizontalalignment','center');
hold on;
axis(1.2*[-1 1 -1 1]);
box on;
drawnow;
energy(data);
```

```
function[s_1,v_1,theta_1,theta_2,w_1,w_2]=db_pendulum
        (theta1,theta2,w1,w2,...
    s,v,mass,mass1, mass2, d1, d2, g, dt)
global t tdata1 fdata vdata theta1data1 theta2data1 sdata1
     s1data s2data s3data mode xend w1data1 w2data1 ueq usw f...
   u1 u2 u3 u4 u5 u6 r1 r2 r3 j F F1 s1 s2 s3 ;
if mode==1
    k1=10;
    k2=10;
    f=k1*(xend-s)-k2*v;

elseif mode==2
    f=ueq+usw;

elseif mode==3
   M=[mass+mass1+mass2 (mass1+mass2)*d1*cos(theta1)
        mass2*d2*cos(theta1);...
   (mass1+mass2)*d1*cos(theta1) (mass1+mass2)*d1^2
        mass2*d1*d2*cos(theta1-theta2);...
   mass2*d2*cos(theta2) mass2*d1*d2*cos(theta1-
theta2)mass2*d2^2];
   C=[0 -1*(mass1+mass2)*d1*w1*sin(theta1) -
1*mass2*d2*w2*sin(theta2);...
   0 0 mass2*d1*d2*w1*sin(theta1-theta2);...
   0 -1*mass2*d1*d2*w1*sin(theta1-theta2) 0];
   G=[0;(mass1+mass2)*g*d1*sin(theta1);mass2*g*d2*sin(theta2)];
   D=2*inv(M);
   bi1=D(1,1);
   bi2=D(2,1);
   bi3=D(3,1);
   B=inv(M)*(-G-C*[v;w1;w2]);
   f1=B(1,1);
   f2=B(2,1);
   f3=B(3,1);
```

```
    K=0.5;ante=0.02;
    a1=10;a2=0.2;a3=-15;
    b1=4.5;b2=0.05;b3=-0.25;
% first_level sliding surface
    s1=v+a1*(s-xend);
    s2=w1+a2*theta1;
    s3=w2+a3*theta2;
% second_level sliding surface
    S=b1*s1+b2*s2+b3*s3;
    ue=-(b1*a1*v+b1*f1+b2*a2*w1+b2*f2+b3*a3*w2+b3*f3)/
        (b1*bi1+b2*bi2+b3*bi3);
    us=-(K*S+ante*sign(S))/(b1*bi1+b2*bi2+b3*bi3);
    f=(ue+us);

elseif mode==4
    f=1.5*(mass+mass1+mass2)*((0.7+0.3*r1)*(u1+u2)+(0.2*r2)
         *(u3+u4)+(1.3+2.2*r3)*(u5+u6));
    a=((mass1+mass2)/mass1)*(1/d1+1/d2);
    b=((mass1+mass2)/mass1)^2*(1/d1+1/d2)^2
      -4*((mass1+mass2)/mass1)/(d1*d2);
    dr=0.01;
    k=exp(-(dr*pi)/((1-dr^2)^0.5));


    if (f-F1)>0.06
         f=F1+0.06;
    elseif (f-F1)<-0.06
         f=F1-0.06;
    end
    F1=f;

elseif mode==6
    k1=10;
    k2=10;
```

```
l1=0.5;
l2=0.5;
R=2; %m2/m1

beta=sqrt((1+R)^2*(1/l1+1/l2)^2-4*((1+R)/l1/l2));

fw2=sqrt(g/2)*sqrt((1+R)*(1/l1+1/l2)-beta); %angular
    frequencies
fw1=sqrt(g/2)*sqrt((1+R)*(1/l1+1/l2)+beta);

ro=pi/fw1;
ro1=pi/fw2;

if t<ro % the convolved input shaper and the passivity-based
        controller
    f=1/16*(k1*(xend-s)-k2*v);
    elseif t>=ro&&t<2*ro
    f=(1/16+1/8)*(k1*(xend-s)-k2*v);
    elseif t>=2*ro&&t<ro1
    f=(1/16+1/8+1/16)*(k1*(xend-s)-k2*v);
    elseif t>=ro1&&t<ro+ro1
    f=(1/16+1/8+1/16+1/8)*(k1*(xend-s)-k2*v);
    elseif t>=ro+ro1&&t<2*ro+ro1
    f=(1/16+1/8+1/16+1/8+1/4)*(k1*(xend-s)-k2*v);
    elseif t>=2*ro+ro1&&t<2*ro1
    f=(1/16+1/8+1/16+1/8+1/4+1/8)*(k1*(xend-s)-k2*v);
    elseif t>=2*ro1&&t<2*ro1+ro
    f=(1/16+1/8+1/16+1/8+1/4+1/8+1/16)*(k1*(xend-s)-k2*v);
    elseif t>=2*ro1+ro&&t<2*ro1+2*ro
    f=(1/16+1/8+1/16+1/8+1/4+1/8+1/16+1/8)*(k1*(xend-s)-k2*v);
    elseif t>=2*ro1+2*ro
    f=(1/16+1/8+1/16+1/8+1/4+1/8+1/16+1/8+1/16)*
      (k1*(xend-s)-k2*v);
end
```

```
elseif mode==7

    k1=10;
    k2=10;

    A1=0.2152;A2=0.2848;A3=0.2848;A4=0.2152;
    t1=0.2382;t2=0.8727;t3=1.2290;
    if t<t1
            f=(A1)*(k1*(xend-s)-k2*v);
        elseif t>=t1&&t<t2
            f=(A1+A2)*(k1*(xend-s)-k2*v);
        elseif t>t2&&t<t3
            f=(A1+A2+A3)*(k1*(xend-s)-k2*v);
        elseif t>=t3
            f=(A1+A2+A3+A4)*(k1*(xend-s)-k2*v);
    end

elseif mode==8

    f=1.5*(mass+mass1+mass2)*((0.7+0.3*r1)*(u1+u2)+(0.2*r2)*
        (u3+u4)+(1.3+2.2*r3)*(u5+u6));

    A1=0.2152;A2=0.2848;A3=0.2848;A4=0.2152;
    t1=0.2382;t2=0.8727;t3=1.2290;
    if t<t1
            f=(A1)*f;
        elseif t>=t1&&t<t2
            f=(A1+A2)*f;
        elseif t>t2&&t<t3
            f=(A1+A2+A3)*f;
        elseif t>=t3
            f=(A1+A2+A3+A4)*f;
    end

end
```

```
% the main function of the system
M=[mass+mass1+mass2 (mass1+mass2)*d1 mass2*d2;...
   (mass1+mass2)*d1 (mass1+mass2)*d1^2 mass2*d1*d2;...
   mass2*d2,mass2*d1*d2,mass2*d2^2];

C=[0 -1*(mass1+mass2)*d1*w1*theta1 -1*mass2*d2*w2*theta2;...
   0 0 mass2*d1*d2*w1*(theta1-theta2);...
   0 -1*mass2*d1*d2*w1*(theta1-theta2) 0];
G=[0;(mass1+mass2)*g*d1*theta1;mass2*g*d2*theta2];
T=[f;0;0];
A=[0 1 0 0 0 0;0 0 ((mass1+mass2)*g+d2*mass2*w1*w2)/mass 0 ...
   -1*d2*mass2*w1*w2/mass 0;0 0 0 1 0 0;0 0 ...
   (-1*(mass+mass1)*(mass1+mass2)*g-d2*mass2*w1*w2)/
       (d1*mass*mass1) ...
   0 (mass*mass2*g+d2*mass2*w1*w2)/(mass*mass1*d1)
       0;0 0 0 0 0 1;...
   0 0 ((mass1+mass2)*g+mass2*d2*w1*w2)/(mass1*d2) 0 ...
   (-1*mass2*d2*w1*w2-(mass1+mass2)*g)/(mass1*d2) 0];
B=[0;1/mass;0;-1/(mass*d1);0;0];
P=[-2,-4,-8,-10,-16,-20];
z=place(A,B,P);
x=inv(M)*(T-G-C*[v;w1;w2]);
A=inv(M);
b1=A(1,1);
b2=A(2,1);
b3=A(3,1);
B=inv(M)*(-G-C*[v;w1;w2]);
f1=B(1,1);
f2=B(2,1);
f3=B(3,1);
s1=z(1)*(-xend+s)+z(2)*v;
s2=z(3)*theta1+s1;
s3=z(4)*w1+s2;
s4=z(5)*theta2+s3;
```

```
s5=z(6)*w2+s4;
ueq=-(z(1)*v+z(2)*f1+z(3)*w1+z(4)*f2+z(5)*w2+z(6)*f3)/
     (z(2)*b1+z(4)*b2+z(6)*b3);
usw=(-0.1*sign(s5)-s5)/(z(2)*b1+z(4)*b2+z(6)*b3);

%SIRMs
a1=theta1/0.5;
b1=w1/1.5;
if a1>-1&&a1<0
    u1=(a1^2+2*a1)/(-2*a1^2-2*a1+1);
elseif a1>0&&a1<1
    u1=(2*a1-a1^2)/(-2*a1^2+2*a1+1);
elseif a1<-1
    u1=-1;
elseif a1==0
    u1=0;
elseif a1>1
    u1=1;
end
if b1>-1&&b1<0
    u2=(b1^2+2*b1)/(-2*b1^2-2*b1+1);
elseif b1>0&&b1<1
    u2=(2*b1-b1^2)/(-2*b1^2+2*b1+1);
elseif b1<-1
    u2=-1;
elseif b1==0
    u2=0;
elseif b1>1
    u2=1;
end
a2=theta2/0.5; b2=w2/1.5; if a2>-1&&a2<0
    u3=(a2^2+2*a2)/(-2*a2^2-2*a2+1);
elseif a2>0&&a2<1
    u3=(2*a2-a2^2)/(-2*a2^2+2*a2+1);
```

```
elseif a2<-1
    u3=-1;
elseif a2==0
    u3=0;
elseif a2>1
    u3=1;
end
if b2>-1&&b2<0
    u4=(b2^2+2*b2)/(-2*b2^2-2*b2+1);
elseif b2>0&&b2<1
    u4=(2*b2-b2^2)/(-2*b2^2+2*b2+1);
elseif b2<-1
    u4=-1;
elseif b2==0
    u4=0;
elseif b2>1
    u4=1;
end
if (xend-s)>=-1&&(xend-s)<0
    u5=((xend-s)^2+2*(xend-s))/(-2*(xend-s)^2-2*(xend-s)+1);
elseif (xend-s)>0&&(xend-s)<=1
    u5=(2*(xend-s)-(xend-s)^2)/(-2*(xend-s)^2+2*(xend-s)+1);
elseif (xend-s)<-1
    u5=-1;
elseif (xend-s)==0
    u5=0;
elseif (xend-s)>1
    u5=1;
end
if v>-1&&v<0
    u6=(-1*v^2-2*v)/(-2*v^2-2*v+1);
elseif v>0&&v<1
    u6=(v^2-2*v)/(-2*v^2+2*v+1);
elseif v<-1
```

```
    u6=1;
elseif v==0
    u6=0;
elseif v>1
    u6=-1;
end
if theta1>-0.5&&theta1<0.5
    r1=(abs(theta1)-abs(theta1)^2)/(0.5+2*abs(theta1)
       -4*abs(theta1)^2);
    r2=1;
elseif abs(theta1)>=0.5&&abs(theta1)<=1
    r1=(5*abs(theta1)-3*abs(theta1)^2-1.5)/(6*abs(theta1)
       -4*abs(theta1)^2-1.5);
    r2=0;
elseif abs(theta1)>1
    r1=1;
    r2=0;
end

   a3=max(abs(a1),abs(a2));
   b3=min(abs(a1),abs(a2));

if abs(a1)>=0.5&&abs(a1)<=1&&abs(a2)>=0.5&&abs(a2)<=1
    r3=0;
elseif a1>-0.5&&a1<0.5&&a2>-0.5&&a2<0.5
    r3=(a3-a3^2+0.5-2*b3^2)/(2*a3-2*a3^2+0.5-2*b3^2);
elseif a1>-0.5&&a1<0.5&&abs(a2)>=0.5&&abs(a2)<=1
    if (0.5-abs(a1))<=(abs(a2)-0.5)
      r3=(abs(a1)-3*a1^2+0.5)/(4*abs(a2)-2*a2^2
         -4*a1^2+2*abs(a1)-1);
   elseif (0.5-abs(a1))>(abs(a2)-0.5)
      r3=(0.5-2*a1^2+abs(a2)-a2^2)/(6*abs(a2)
         -4*a2^2-2*a1^2-1);
   end
```

```
elseif a2>-0.5&&a2<0.5&&abs(a1)>=0.5&&abs(a1)<=1
   if (0.5-abs(a2))<=(abs(a1)-0.5)
      r3=(abs(a2)-3*a2^2+0.5)/(4*abs(a1)-2*a1^2
         -4*a2^2+2*abs(a2)-1);
   elseif (0.5-abs(a2))>(abs(a1)-0.5)
      r3=(0.5-2*a2^2+abs(a1)-a1^2)/(6*abs(a1)-4*a1^2-2*a2^2-1);
   end
elseif abs(a1)>1||abs(a2)>1
   r3=0;
end
y=[v;w1;w2]+dt*x;
v_1=y(1);
w_1=y(2);
w_2=y(3);
z=[s;theta1;theta2]+0.5*dt^2*x+dt*[v;w1;w2];
s_1=z(1);
theta_1=z(2);
theta_2=z(3);
t = t + dt;
tdata1 = [tdata1,t];
fdata = [vdata,f];
vdata = [vdata,v_1];
theta1data1=[theta1data1,theta_1];
theta2data1=[theta2data1,theta_2];
w1data1=[w1data1,w_1];
w2data1=[w2data1,w_2];
sdata1=[sdata1,s_1];
s1data=[s1data,s1];
s2data=[s2data,s2];
s3data=[s3data,s3];

function [s_1,v_1,theta_1,theta_2,w_1,w_2,x_1]=Endb_pendulum
     (theta1, theta2, w1, w2,...
   s,v, g, dt,x_1)
```

```
global t tdata1 fdata vdata theta1data1 theta2data1 xend w1data1
     w2data1 sdata1 f;
 len1=3;
 len2=2;
 m=10;
 m1=25;
 m2=40;
 a=-0.01;
 x0=x_1;
 b=(m1+m2)*len1/(m2*len2);
 e=s-xend+a*(b*sin(theta1)+sin(theta2));% the composite signal

 kd=120;
 tspan=[0:0.001:0.01];
 [ts,ix]=ode45(@(ts,x)dxdt(ts,x,e),tspan,x0);
 x=ix(8,:);
 x_1=x;

 kp=40;
 f=-2*kp*atan(e)/pi-2*kd*atan(x+kd*e)/pi; % control force
 M=[m+m1+m2  (m1+m2)*len1 m2*len2;...
  (m1+m2)*len1  (m1+m2)*len1^2 m2*len1*len2;...
  m2*len2,m2*len1*len2,m2*len2^2];
 C=[0 -1*(m1+m2)*len1*w1*theta1 -1*m2*len2*w2*theta2;...
  0 0 m2*len1*len2*w1*(theta1-theta2);...
  0 -1*m2*len1*len2*w1*(theta1-theta2) 0];
 G=[0;(m1+m2)*g*len1*theta1;m2*g*len2*theta2];
 T=[f;0;0];
 x=inv(M)*(T-G-C*[v;w1;w2]);
 y=[v;w1;w2]+dt*x;
v_1=y(1);
w_1=y(2);
w_2=y(3);
z=[s;theta1;theta2]+0.5*dt^2*x+dt*[v;w1;w2];
```

```
s_1=z(1);
theta_1=z(2);
theta_2=z(3);
t = t + dt;
tdata1 = [tdata1,t];
fdata = [vdata,f];
vdata = [vdata,v_1];
theta1data1=[theta1data1,theta_1];
theta2data1=[theta2data1,theta_2];
w1data1=[w1data1,w_1];
w2data1=[w2data1,w_2];
sdata1=[sdata1,s_1];

% ------In the mode==5 Auxiliary function
function dx=dxdt(ts,x,e)
dx=-120*x-120^2*e;
%
% function [s_1,v_1,theta_1,theta_2,w_1,w_2]=Edb_pendulum
     (theta1, theta2, w1, w2,...
%   s,v,mass,mass1, mass2, d1, d2, g, dt)
%
% end
% --- Executes on slider movement.
function gravity_Callback(hObject, eventdata, handles)
% hObject    handle to gravity (see GCBO)
% eventdata  reserved - to be defined in a future
     version of MATLAB
% handles    structure with handles and user data (see GUIDATA)
handles.G=get(hObject,'value');
set(handles.gr,'string',num2str(handles.G));
guidata(hObject, handles);

% Hints: get(hObject,'Value') returns position of slider
%     get(hObject,'Min') and get(hObject,'Max') to
           determine range of slider
```

```
% --- Executes during object creation, after
         setting all properties.
function gravity_CreateFcn(hObject, eventdata, handles)
% hObject    handle to gravity (see GCBO)
% eventdata  reserved - to be defined in a
     future version of MATLAB
% handles    empty - handles not created until after all
     CreateFcns called
% Hint: slider controls usually have a light gray background.
if isequal(get(hObject,'BackgroundColor'),
          get(0,'defaultUicontrolBackgroundColor'))
   set(hObject,'BackgroundColor',[.9 .9 .9]);
end

function energy(data)
 Epot1=-data.mass1*data.g*data.len1*(cos(data.t1)-1.000);
 Epot2=-data.mass2*data.g*((cos(data.t1)-1.000)*data.len1+...
   (cos(data.t2)-1.000)*data.len2);
 Ekin1=0.5*data.mass1*(data.w1*data.len1).^2;
 Ekin2=0.5*data.mass2*( (data.w2*data.len2*cos(data.t2) +...
   data.w1*data.len1*cos(data.t1)).^2+(data.w2*data.len2*
       sin(data.t2)...+data.w1*data.len1*sin(data.t1)).^2 );
 Etot=Ekin1+Ekin2+Epot1+Epot2;

 set(data.U1,'string',num2str(Epot1));
 set(data.U2,'string',num2str(Epot2));
 set(data.T1,'string',num2str(Ekin1));
 set(data.T2,'string',num2str(Ekin2));
 set(data.totE,'string',num2str(Etot));

% --- Executes on selection change in popupmenu1.
function popupmenu1_Callback(hObject, eventdata, handles)
% hObject    handle to popupmenu1 (see GCBO)
```

```
% eventdata  reserved - to be defined in a future
     version of MATLAB
% handles    structure with handles and user data (see GUIDATA)

% Hints: contents = cellstr(get(hObject,'String')) returns
        popupmenu1 contents as cell array
%         contents{get(hObject,'Value')} returns
        selected item from popupmenu1 global mode;
mode=get(handles.popupmenu1,'Value');

% --- Executes during object creation, after setting all
     properties.
function popupmenu1_CreateFcn(hObject, eventdata, handles)
% hObject    handle to popupmenu1 (see GCBO)
% eventdata  reserved - to be defined in a future version of
     MATLAB
% handles    empty - handles not created until after all
     CreateFcns called
% Hint: popupmenu controls usually have a white background on
     Windows.
%        See ISPC and COMPUTER.
if ispc && isequal(get(hObject,'BackgroundColor'),
        get(0,'defaultUicontrolBackgroundColor'))
   set(hObject,'BackgroundColor','white');
end

function send_Callback(hObject, eventdata, handles)
% hObject    handle to send (see GCBO)
% eventdata  reserved - to be defined in a future version of
     MATLAB
% handles    structure with handles and user data (see GUIDATA)
global xend;
```

```
xend=str2double(get(hObject,'String'));
if xend~=1
   errordlg('Not an expected position','Please enter 1');
else guidata(hObject, handles);
end

% Hints: get(hObject,'String') returns contents of send as text
%        str2double(get(hObject,'String')) returns contents of
     send as a double

% --- Executes during object creation, after setting all
     properties.
function send_CreateFcn(hObject, eventdata, handles)
% hObject    handle to send (see GCBO)
% eventdata  reserved - to be defined in a future version of
     MATLAB
% handles    empty - handles not created until after all
     CreateFcns called

% Hint: edit controls usually have a white background on
     Windows.
%       See ISPC and COMPUTER.
if ispc && isequal(get(hObject,'BackgroundColor'),
     get(0,'defaultUicontrolBackgroundColor'))
   set(hObject,'BackgroundColor','white');
end

% --- Executes on button press in info.
function info_Callback(hObject, eventdata, handles)
% hObject    handle to info (see GCBO)
% eventdata  reserved - to be defined in a future version of
     MATLAB
% handles    structure with handles and user data (see GUIDATA)
```

```
msgbox({'School of Control and Computer Engineering';'North
     China Electric Power Univeristy, Beijing';'V1.0';'by
     Dianwei Qian, Binbin Yang and Qiang Wang'},'All rights
     reserved');

% --- Executes on key release with focus on figure1 and
     none of its controls.
function figure1_KeyReleaseFcn(hObject, eventdata, handles)
% hObject    handle to figure1 (see GCBO)
% eventdata  structure with the following fields (see FIGURE)
% Key: name of the key that was released, in lower case
% Character: character interpretation of the key(s) that was
     released
% Modifier: name(s) of the modifier key(s) (i.e., con-
trol, shift) released
% handles    structure with handles and user data (see GUIDATA)
```

Index

DOI 10.1515/9783110520101-010

Bei Fragen zur Produktsicherheit wenden Sie sich bitte an:
If you have any questions regarding product safety,
please contact:

Walter de Gruyter GmbH
Genthiner Straße 13
10785 Berlin
productsafety@degruyterbrill.com